# THE HIDDEN COLORS OF TINY LIFE: DIVING INTO MICROBIAL PIGMENTS

# FROM AND TO ALLAH

# Authors

**Dr. Tarek H. Taha** is an Associate Professor at the Environmental Biotechnology Department, Genetic Engineering and Biotechnology Research Institute (GEBRI), City of Scientific Research and Technological Applications (SRTA-CITY), Alexandria, Egypt. He was a visiting Professor at Newcastle University, UK. He has his expertise in the field of Environmental Biotechnology. His research interest is concerned by the Biomonitoring and Bioremediation of environmental contaminants. He is also interested in the biosynthesis of nanoparticles and their applications in biosensors and other environmental fields, and has a great passion with Bioinformatics, Molecular techniques, and Genetic engineering. In addition, he is interested in the production of biofuel from environmental wastes, and finally, the bioconversion of environmental wastes into industrial and pharmaceutical products.

**Dr. Ghada E. Hegazy** is a researcher at Marine Microbiology Department, National Institute of Oceanography and Fisheries (NIOF), Egypt. Her research focuses on the isolation of halophilic archaea and study their applications in different aspects. Dr. Hegazy has an outstanding experience in Microbial Fuel cell synthesis for energy production, using marine archaea. She is interested in the nanoparticle's biosynthesis and their applications in biosensor field, molecular technique, biosorption of heavy metals and bioremediation of organic contaminants in the marine environment, and finally, she succeeds in the production of carotenoids and bio-surfactants from archaea with anticancer and antiviral activity against HCV, HBV, ADV, HSV and SARS viruses.

**Dr. Marwa M. Eltarahony** is an associate professor at Environmental biotechnology department, Genetic Engineering and Biotechnology Research Institute (GEBRI), City of Scientific Research and Technological Applications (SRTA-City), Egypt. Her researches focus on biological synthesis of nanoparticles/ nanocomposites and their application in different sectors. She is interested in employing novel materials as anticancer and biocide agents. Besides, she participated in bio-remediating the environment from different organic and inorganic pollutants using different hybrid techniques, and finally she succeeded in extraction of different microbial secondary metabolites and recruiting them in water treatment.

**Dr. Amira Sabry** is a Researcher at the Protein Research Department, Genetic Engineering & Biotechnology Research Institute, City of Scientific Research and Technological Applications (Alexandria, Egypt). She is interested in the production and purification of recombinant therapeutic proteins and evaluating their biological activities. Also, in the enzymes production and studying their biotechnological applications. In addition, investigating the virulence factors contributing to the pathogenicity of the microorganisms. Furthermore, she is interested in proteomics, molecular biology and large-scale production of biosurfactants. Dr. Amira Sabry has succeeded in the expression of soluble and active human tumor necrosis factor-alpha (hTNF-α) by an *in vitro* method in an *E. coli*-based cell-free protein synthesis (CFPS) system and assessing its anticancer activity against human cancer cell lines.

**Dr. Khouloud M.I. Barakat** is Professor at Marine Microbiology Department, National Institute of Oceanography and Fisheries (NIOF), Egypt. Her expertise is biotechnological studies on marine bacteria, yeast and fungi. Her studies are focused on the marine bacteria and fungi resources in different habitats. Her researches were carried on antibiotic resistance in marine microbes and antibiotic structure-activity relationships. Marine natural products and antibiotic synergism are her main target. Fish microbiology and chemotherapy for fish pathogenicity were discussed in many of her researches and published books.

**Dr. Hussein Oraby** is a researcher at Chemical Engineering Department, Military Technical College (MTC), Egypt. His research focuses on the polymer composites processing and their applications in the fields of membrane production (water treatment and desalination) and electromagnetic interference shielding applications (EMI SE).

## Contents

### CHAPTER I
# Assets in The Bio-Pigments Realm and Their Various Applications — 1

| | |
|---|---|
| Abstract | 1 |
| Keywords | 2 |
| Background | 2 |
|    1. Types of bacterial pigments | 6 |
|    1.1. Tetrapyrrole derivatives: chlorophylls and heme colors | 6 |
|    1.2. Isoprenoid derivatives: carotenoids and iridoids | 8 |
|    1.3. N-heterocyclic compounds | 9 |
|    1.4. Benzopyran derivatives (oxygenated heterocyclic substances): anthocyanins and alternative flavonoid pigments | 10 |
|    1.5. Polyketide | 11 |
|    1.6. Melanins | 11 |
|    2. Pigments recovery | 12 |
|    3. Multifaceted applications of bacterial pigments | 17 |
| a. Bacterial pigments in red biotechnology | 19 |
| i. Antibacterial pigments | 20 |
| ii. Antifungal pigments | 22 |
| iii. Antiviral pigments | 24 |
| iv. Anticancer / Antineoplastic pigments | 25 |

| | |
|---|---|
| v. Anti-inflammatory pigments | 26 |
| vi. Antioxidant Activity | 26 |
| b. Bacterial pigments and white biotechnology | 27 |
| i. Solar cells and microbial fuel cells | 27 |
| ii. Cosmetics | 28 |
| ii. Textile dying | 30 |
| iv. Candle and paper industry | 32 |
| 4. Conclusion and future perspectives | 34 |
| References | 36 |

## CHAPTER II

# Biopigments From Actinomycetes and Related Microbes: Applications and Diverse Activities
45

| | |
|---|---|
| Abstract | 45 |
| Keywords | 45 |
| Introduction | 46 |
| Production of natural pigments by actinomycetes | 49 |
| Industrial applications of pigments | 51 |
| - As food colorants | 51 |
| - As dying agents | 52 |
| - As cosmetics ingredient | 53 |
| - As antifouling agent | 53 |
| Types of biopigments activities | 55 |
| - Pharmacological activity | 55 |
| - Antimicrobial activity | 56 |
| - Anticancer activity | 58 |
| - Antioxidant, Anti-allergic and anti-inflammatory activities | 59 |
| Stability of biopigments | 60 |

x

| | |
|---|---|
| - Encapsulation | 60 |
| - Nano-emulsion | 62 |
| Conclusion | 62 |
| References | 63 |

## CHAPTER III

## Archaeal Pigments: Synthesis Mechanisms and Multifaceted Applications — 76

| | |
|---|---|
| Abstract | 76 |
| Keywords | 77 |
| Introduction | 77 |
| - Types, contents, and biosynthesis of halophilic carotenoids | 81 |
| - Potential of carotenoid production from halophilic archaea | 87 |
| Carotenoids Biological activities | 89 |
| - Bacterioruberin as antioxidant | 90 |
| - Membrane rigidity is controlled by bacterioruberin | 91 |
| - Bacterioruberin part of the rhodopsin complex | 91 |
| Biotechnological applications of halophilic archaea carotenoids | 92 |
| Conclusion | 94 |
| References | 95 |

## CHAPTER IV

## Fungal Pigments: Biosynthesis Pathways and Versatile Applications — 101

| | |
|---|---|
| Abstract | 101 |

| | |
|---|---|
| Keywords | 102 |
| Introduction | 102 |
| Fungal carotenoids classification | 103 |
| Major carotenoids producing fungi | 103 |
| - β-Carotene | 105 |
| - Astaxanthin | 106 |
| - Torulene | 106 |
| - Torularhodin | 107 |
| Fungal carotenoids biosynthetic pathway | 107 |
| Carotenoids located within the mould cells | 119 |
| Fungal carotenoids therapy | 120 |
| Fungal carotenoids industry | 123 |
| Metabolic engineering | 127 |
| Conclusion | 128 |
| References | 129 |

## CHAPTER V

## Microalgae-Derived Pigments: An Introduction to their Biosynthesis, and Applications

| | |
|---|---|
| Abstract | 142 |
| Keywords | 143 |
| 1. Introduction | 143 |
| 2. Microalgae-derived pigments | 145 |
| 2.1. Chlorophylls | 146 |
| 2.2. Carotenoids | 148 |
| 2.3. Phycobiliproteins (PBPs) | 151 |
| 3. Factors affecting the microalgal pigment production | 155 |
| 3.1. Light | 156 |
| 3.2. Temperature | 158 |

| | |
|---|---|
| 3.3. Culture media | 159 |
| 3.3.1. Nitrogen | 159 |
| 3.3.2. pH and salinity | 160 |
| 3.3.3. Micronutrients | 162 |
| 4. 4. Potential applications of microalgal pigments | 163 |
| 5. Future perspective and conclusive remarks | 171 |
| References | 172 |

## CHAPTER VI

# Analytical Insights into Natural Marine Organic Matter: Compositional and Structural Characteristics — 180

| | |
|---|---|
| Abstract | 180 |
| Keywords | 180 |
| 1. Introduction | 181 |
| 2. Extraction, Separation, and Purification Techniques for Organic Matter Isolation from Seawater and Sediments | 182 |
| 3. Carbohydrate analysis | 185 |
| 3.1 Total carbohydrate analysis | 185 |
| 3.2 Analysis of monosaccharide composition | 190 |
| 4. Lipid analysis | 192 |
| 4.1 Chlorophyll pigment analysis | 194 |
| 4.2 Hydrocarbon analysis | 195 |
| 4.3 Fatty acids and sterols | 197 |
| 4.4 Phospholipids | 199 |
| 5. Protein analysis | 200 |
| 6. Polyphenolic and lignin analysis | 204 |
| 7. Structural characteristics of OM | 205 |
| 7.1 OM formation and aggregation | 205 |

| | |
|---|---|
| 7.2 Spectroscopic and chromatographic methods for OM structural characterization | 207 |
| 8. Conclusion | 209 |
| References | 210 |

# CHAPTER I

## Assets in The Bio-Pigments Realm and Their Various Applications

**Marwa Eltarahony,** *PhD*

Environmental Biotechnology Department, Genetic Engineering and Biotechnology Research Institute (GEBRI), City of Scientific Research and Technological Applications (SRTA-City), New Borg El-Arab City 21934, Alexandria, Egypt.

### Abstract

As modern Phoenicians, microbiologists, and biotechnologists pay their attention to plentiful sources of microbial pigments, especially bacteria. The short life-cycle, superior productivity, wide strain versatility, availability for cultivation regardless of seasonal variation, and ubiquitous in extreme ecosystems of bacteria are the major reasons for their preference for other bio-pigments. Indeed, numerous bacterial genera produce various chemically different colored compounds that exhibit diverse physiological functions. In this scenario, to commercialize the production process of bacterial pigments in a low-cost technology, various aspects were taken into consideration. Wherein, process optimization, utilizing gene manipulation approaches and employment of bioreactors or fermenters with agro-industrial residues, poultry waste, and fish waste would enhance productivity and simultaneously follow the principles of the circular economy. The synergism of all previous tools would undoubtedly fulfill the industrial needs in a sustainable, environmentally friendly, and economic manner with zero-pollution output. Remarkably, the advancement in immobilization techniques to formulate pigment in microspheres or nano-capsules to maintain stabilization against different operation conditions is crucially required. Ultimately, bacterial pigments displayed promising multifaceted performance in a wide range of applications under the umbrella of red, yellow, white, and green biotechnology.

**Keywords:** Bacterial pigments, Pigment production, Biosynthesis, Natural pigments, Photoprotection

## Background

The worldwide developmental process influenced all fields of life by providing rapidity, efficacy, and comfort. Therefore, synthetic dyes have vastly overtaken the coloration of several materials beginning from textile fibers and paper; passing through tannery, leather, and cosmetics; and ending with food and pharmaceutical products. Their ease of use, lower cost, and their wide range of chemical structures boosted their broad applications (Slama et al., 2021). Meanwhile, some synthetic coloring agents, which were originally approved by the Food and Drug Administration (FDA) for usage, were later found to promote cancer and hyperactivity in children. Besides, sunset yellow and tartrazine result in allergic effects, benzidine dyes result in bowel cancer, and carbon black, widely used as printing ink, is also a potential carcinogen; thus, had been withdrawn from use due to their apparent hazards (Kumar et al., 2022; Gomes and Soares, 2023).

In addition, unethical and untreated discharge of industrial dye effluents produces toxic compounds and persists longer in the environment. Knowing the drawbacks of these synthetic coloring agents, consumers in recent years have become increasingly aware of the artificial coloring agents being added to their food; thus, require their food to be as "natural" as possible. Hence, an urgent requirement to replace synthetic dyes with natural ones, in particular for human consumption in food or medicine as additives, as they are biodegradable and nontoxic to humans (Azman et al., 2018). Natural pigments or bio pigments are steamed from ores, insects, plants, and microbes. However, their production from microorganisms, as secondary metabolites, is preferred over those from plants by the dint of their stability and availability for cultivation throughout the year. Among microbes, bacteria have immense potential to produce diverse kinds of pigments, which find their way into various sectors but in limited industrial applications. Wherein, most of the bacterial pigment production is still at the R&D stage. Interestingly, the easy propagation of bacteria, short life-cycle, simplicity/fast culturing techniques, their wide strain selection (i.e., versatility), high productivity over other sources, and easiness in gene manipulation are the major reasons that attract scientists and biotechnologists to investigate bacterial pigments (Ramesh et al., 2022). Let alone the feasibility of bulk production on cheap and sustainable substrates such as agro-industrial waste material; allowing by such way continuous bioreactor operation and cost-effective fermentation systems that are adequate for industrial production in less-complex processes with zero-pollution output (Usman et al., 2017). Moreover, their natural features, safety, eco–friendly, biocompatibility, medicinal properties, nutritional value as vitamins, stability (i.e., heat, light exposure, and pH variation), high water solubility, and their production is independent of season or geographical conditions with controllable and predictable yield elevate the need for high yield productivity (Venil et al., 2013). Additionally, the simplicity of the extraction approach (i.e., liquid–liquid extraction technique) and the possibility of recovering the used solvent for subsequent use would minimize the operating cost; hereby, appealing to the market makers for commercial production (Ahmad et al., 2012; Vinotha et al., 2019). Remarkably, the United States Food and Drug Administration (US FDA) and European Food Safety Authority (EFSA) have approved a few bacterial pigments for the nutraceutical industry, such as astaxanthin and β–carotene from phototrophic bacteria (Orlandi et al., 2022).

It is worth mentioning that various bacterial species produce different shades of color, and are ubiquitous in several ecological areas, from tropical to Polar Regions and from deep-sea to aerial districts. They are found not only in marine, sediments, and different soil environments but can also be found in extreme environments such as desert sands, ice cores, glaciers, air–water interfaces, underlying waters, deep sea hydrothermal vents, hot springs, lava caves, salt regions. Intriguingly, pigment-producing bacteria were detected in extreme regions such as Himalaya and Antarctic as referred to by Kumar et al., (2022). Moreover, pigments-producing bacteria were reported to be associated with invertebrates and also isolated from organic residues, spoiled fruits and vegetables, damping sites and industrial effluent, endophytes, and dried seafood (Narsing et al., 2017; Azman et al., 2018). Their presence in extreme conditions, either as true or adaptive-dwellers, triggers them to acclimatize to such ecosystems and ultimately produce novel chromogenic compounds with unique properties that suit industrial and biotechnological applications (Hassan et al., 2022).

## 1. Types of bacterial pigments

The vast array of microbial groups is characterized by their ability to produce various types of pigments as tabulated in Table 1. According to that reported in Sánchez-Muñoz et al., (2020), the bacterial pigments can be classified according to their structural features as listed below, and some of them are illustrated in Figure 1 (Ramesh et al., 2022).

### 1.1. Tetrapyrrole derivatives: chlorophylls and heme colors

Tetrapyrrole derivatives have pyrrole rings in linear or cyclic arrays. Pyrrole is a colorless liquid substance, soluble in alcohol, ether, diluted acids, and also in most organic chemicals. It is composed of four pyrrole rings connected by a methane bridge. They constitute the heme group, which is part of chlorophyll and billin pigments. The functions of tetrapyrroles in microorganisms include light capture and electron transfer reactions. They are cofactors for essential enzymes and sensory proteins. Thus, tetrapyrroles can contribute to oxidative stress, protect cells in photosynthetic bacteria, and contribute to the detoxification of ROS. Tetrapyrrole is biosynthesized from γ-aminolevulinic acid (ALA), a five-carbon amino acid formed by the condensation of glycine and succinyl coenzyme A (C4 pathway) or from α-ketoglutarate (C5 pathway). There are two groups of tetrapyrrole pigments, chlorophylls, and phycobiliproteins. Chlorophylls are the most abundant pigments on earth, and are essential for photosynthesis, absorbing light and transducing it into chemical energy. Their structure contains a hydrocarbon tail connected to chlorine, an aromatic ring with tetrapyrrole. The magnesium ion in the center is bound to four pyrrole rings. These pigments absorb light in the violet-blue (400–500 nm) and the yellow-orange/red (600–700 nm) parts of the visible spectrum. They differ in the degree of unsaturation of the pyrrolic macrocycle in the side chains, influencing their light absorption properties. Chlorophyll a, d, and f are present in cyanobacteria. It is a flammable liquid with eminent biological activities and its derivatives have been found effective against dengue vectors (Sánchez-Muñoz et al., 2020).

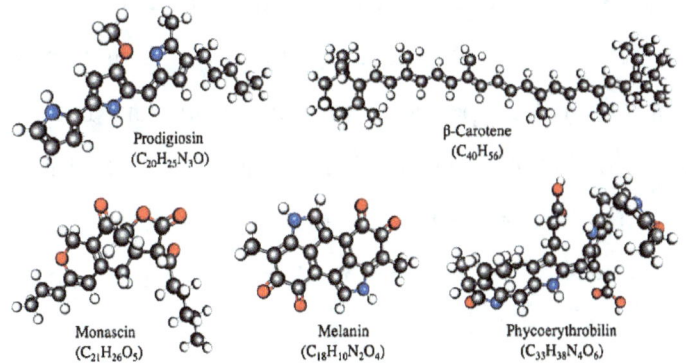

**Figure 1** 3D chemical structure of industrially important bio- pigments (Ramesh et al., 2022).

## 1.2. Isoprenoid derivatives: carotenoids and iridoids

In general, carotenoids have eight isoprenoid units whose order is inverted at the molecule center. Generally, there are over 600 known carotenoids, which are divided into two classes: xanthophylls (which contain oxygen) and carotenes (which are purely hydrocarbons and contain no oxygen). Among the various carotenoids, the most important carotenoids are alpha and beta-carotenes, cryptoxanthin, lutein, lycopene, violaxanthin, neoxanthin, zeaxanthin, and canthxanthin. All carotenoids can be considered as lycopene ($C_{40}H_{56}$) derivatives by reactions adding or removing hydrogen, inserting oxygen, of cyclization, of double bond or methyl migration, of chain elongation and shortening. Lycopene is a solid red bright pigment, can absorb light in the range of 450-570 nm, used as colorant in food and to stabilize flavor formulation, with antioxidant, anti-inflammatory, and neuroprotective properties. The most prominent function of carotenoids is their contribution to harvest light energy. They absorb light and pass the excitation energy onto chlorophyll, thereby extending the wavelength range of harvested light. Several bacterial groups produce a considerable amount of carotenoids. β- Carotene can be a red, brownish-red or purple-violet crystal or powder, depending on the solvent used for extraction and conditions of crystallization. Carotenoids are soluble in benzene, chloroform, and carbon disulfide; moderately soluble in ether, petroleum ether and oils. β - Carotene is approved by U.S. Food and Drug Association to be used as a nutrient supplement and as a source of vitamin A and also for obesity treatment. More recently they have been used for nutraceutical, cosmetic, and pharmaceutical purposes. While the direct exposure of carotenoids can cause eye and skin irritation and harmful to aquatic life with long lasting effects (Sánchez-Muñoz et al., 2020; Barreto et al., 2023).

### 1.3. N-heterocyclic compounds

There are different from tetrapyrroles: pterins, purines, flavins, phenoxazines, phenazines and betalains. In flavins, a pteridine and a benzene ring are condensed. Riboflavin, or vitamin B2 is a solid orange-yellow pigment, soluble in water, sodium chloride solution and dilute alkali solution; consisting of a ribose sugar unit and a three-ring flavin structure known as a lumichrome. Based on the structure, they can be divided into the yellow betaxanthins (480 nm) and red-purple betacyanins (540 nm) and their color is due to the resonating double bonds. A riboflavin analog produced by *Streptomyces* bacteria. Generally, riboflavin is biosynthesized using the purine guanosine triphosphate (GTP) and ribulose-5-phosphate (Ru5P) from the pentose phosphate pathway as precursors. For over 30 years, riboflavin supplements have been used as part of the phototherapy treatment for neonatal jaundice. Riboflavin co-treatment with b blockers showed improvement against migraine headaches. Riboflavin in combination with UV light has been shown to be effective in reducing harmful pathogens found in blood products (Narsing et al., 2017). Deazaflavins, on the other hand, have diaminouracil and tyrosine as precursors, and their biosynthetic pathway was demonstrated to involve the participation of 50- deoxyadenosyl radicals to form the typical heterocyclic ring. These flavin pigments are fundamental mediators between two-electron and one electron processes in biological systems (Barreto et al., 2023).

### 1.4. Benzopyran derivatives (oxygenated heterocyclic substances): anthocyanins and alternative flavonoid pigments

Among the flavonoids, the anthocyanins are the most relevant pigments. Chemically, they have 15 carbons with a chrome ring bearing a second aromatic ring B in position 2 (C6-C3-C6) and with sugar molecules bonded at different hydroxylated positions of the primary structure, which gives to them several possibilities of combinations and production of colors.

## 1.5. Polyketide

Polyketides are a structurally diverse class of compounds formed by the repeated condensation of malonyl-CoA derivatives (acetate/malonate pathway) in a process catalyzed by the enzymatic apparatus denominated polyketide synthases (PKSs). Quinones are the major constituents in this group. They range from yellow to red; have the largest structural variation among other groups. The primary structure is a desaturated cyclic ketone, derived from an aromatic mono or polycyclic compound. The acetate/ mevalonate or shikimate pathways can form them from biosynthesized phenolic systems. The former involves condensing acetyl-CoA/malonyl-CoA units to form a polyketide chain subject to appropriate folding and cyclization. Menaquinones and anthraquinones are found in mainly bacteria. Anthraquinones have an anthracene ring with a keto group on position 9, 10 and different functional groups such as -OH, -CH3-OCH3, -CH2OH, -CHO, -COOH at various positions (Sánchez-Muñoz et al., 2020).

## 1.6. Melanins

Melanins are nitrogenous polymeric compounds with the indole ring as monomer and classified as eumelanins, pheomelanins, and allomelanins. In general, melanins are a mixture of macromolecules, characterized by molecular irregularity, photoconductivity and insolubility at most used solvents with brown or black colour. They are biosynthesized from tyrosine with the participation of the enzyme tyrosinase and have 5,6-dihydroxyindole as the main building block. Melanin confers resistance to UV light by absorbing a broad range of the electromagnetic spectrum and preventing photoinduced damage. Melanin is used for mimicry, and protects against high temperatures and chemical stresses. While bacterial melanin genes have been used as reporter genes to screen recombinant bacterial strains. It has anti- HIV properties and is useful for photo voltage generation and fluorescence studies. Melanin is also used to generate monoclonal antibodies for the treatment of human metastatic melanoma. Furthermore, a recent work described the remarkable production of melanins from bacteria, proposing that they are potential sources for industrial melanin production. However, bioprospecting for melanin-producing bacteria, including those associated with other organisms or living in extreme environments, could help find melanin hyperproducers or extremozymes that could increase melanin yields (Barreto et al., 2023).

2. **Pigments recovery**

Downstream engineering processes, including bioproduct extraction, usually comprise more than 50% of the total cost of bioprocesses. Moreover, the cleaner and "greener" recovery processes are regarding environmental subjects, the better for a promising industrial application. The recovery of intracellular biopigments involves essentially two steps: cell disruption and pigments extraction. Cell disruption by mechanical methods, such as bead beater and biomass maceration, are traditionally the mostly used, nonetheless it is considered that chemical methods are more efficient since they are fast and efficient. Generally, chemical driven biopigments extractions are simple, low cost and fast procedures methods. In extraction processes, several factors may influence the microbial pigment yield obtainment, such as temperature, solvent choice, type of pigment, complexity of microbial cell wall, and presence of culture media residues. Conventional extraction of different pigment groups may include maceration, distillations, Soxhlet extraction, and water/solvent infusions (Figure 2).

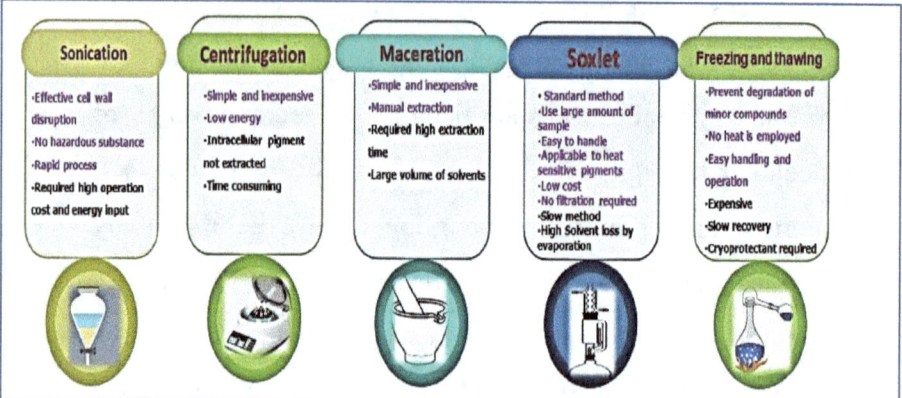

Figure 2 Different conventional methods of extracting bacterial pigments (Rajendranet al., 2023).

Even being used until present days, it is well-known that these methods present disadvantages, such as large volumes of solvent usage, poor extraction, time-consuming processes, chemical degradation of pigment structure when performed with high temperatures, and thus loss of bioactivity. Moreover, during these processes, the use of organic solvents such as acetone, benzene, petroleum ether, hexane, and methanol is prevalent. These solvents are toxic, highly inflammable, and non-biodegradable. Hence, due to sustainability and green technology approaches, pigment extraction without using organic solvents other than ethanol or water is undesirable (Barreto et al., 2023). Regarding extraction methods, the choice for a proper approach involves not only the assessment of the final yield in each procedure, but it is also a reflection about the extract composition. It is known that different extraction procedure will result in distinctive biopigments profile. It is worth noting that intracellular biopigment extraction is still a challenge, since there are no standard and economic procedures for the extraction of biopigments. Another crucial point is the necessity to determine the molecular characteristics of the pigment in order to choose the best solvent or solvents combination for its extraction. Concerning solvent solubility, carotenoids, for instance, are stored in lipid vesicles and therefore can be extracted using various organic non-polar solvents. Otherwise, prodigiosin is a pigment with higher hydrophobicity than many carotenoids and, for this reason, can be easily extracted with acetone. On the other hand, extracellular pigments, cell disruption is not required, but an extraction for the supernatant or from the interstices of microbial biomass may be executed. Extracellular melanin can be precipitated with acetic acid or flocculated with alum (Sánchez-Muñoz et al., 2020). Thence, advanced techniques were developed to extract pigments with minimum environmental hazards compared to conventional methods. Extraction techniques such as electric-pulsed, supercritical fluid extraction (SFE), pressurized liquid extraction (PLE), enzymatic, ultrasound, and microwave-assisted, among others, are described as green newer and feasible methods for obtaining melanin, carotenoids, and chlorophylls. However, solvents such as: ionic liquids (ILs), water, ethanol, esters of fatty acid, essential oils and glycerol etc. are all gaining attention as greener extraction techniques. These techniques usually require less procedure time, with low to average consummation of solvents (water, aqueous, and non-aqueous), protect pigments from degradation, and enhance the quality of natural colorants, in addition to being highly ecofriendly, efficient and financially viable for industrial-scale extraction. Whereas, they may need high pressure and temperature during the process (Sharma et al., 2021; Barreto et al., 2023).

In order to apply the bio-pigments in several application sectors, they should be stable enough. Wherein, stability is one of the vital aspects to be considered for the utilization of natural pigments in industrial formulations. The strength and stability of natural pigments are affected by several factors during processing and storage. Notably, the pigment extracted from microbes, plant resources or agri-food wastes are highly unstable and are susceptible to degradation by external (processing conditions such as temperature, light, oxygen, metals, acids, moisture, and pH) as well as internal factors (e.g. concentration of pigment), in any stage of processing or storage for application. Besides, the bio- pigments are prone to degrade easily in aqueous and lipophilic solutions (Sharma et al., 2021). This degradation occurs through chemical alterations (oxidation, isomerization) in pigment structure, which may result in loss of the bioactivity properties, color fading, release of undesirable odors, limited shelf-life, and thus as a consequence impacting negatively on the quality of the final product. While the stability of these pigments can be increased by various concentrations in the system; low pH and temperature; presence of stabilizers (e.g. chelating agents, antioxidants); absence of light and the presence of acylation or glycosylation in the structure. In this way, formulating bio-pigments in stable formula that enable stabilized application is urgently required as would be discussed in the following section (Venil et al., 2020).

## 3. Multifaceted applications of bacterial pigments

A multidisciplinary scientific model is essential for the development of advanced biotechnology, whose progress is greatly influencing the global economy and the lifestyle of human beings. As defined by United Nation Convention on Biological Diversity, the biotechnology is any technological application that uses biological systems, living organisms, or derivatives thereof, to make or modify products or processes for particular use." Bacterial pigments is falling within the circle of this definition. A plethora of studies and reviews documented the great potential of bacterial pigments in different applicative biotechnological fields.

An overview of the possible uses of pigments can be demonstrated with the biotechnology "rainbow code," which symbolizes different biotechnology branches with different colors. Thus, "red biotechnology" is for human health applications, "yellow biotechnology" is for food and nutrition, "white" indicates industrial biotechnology, and "green biotechnology" indicates to agriculture, plant, and environment. A compendium of the multifaceted applications regarding bacterial pigments in the various fields is depicted in Figure 3 (Venil et al., 2020).

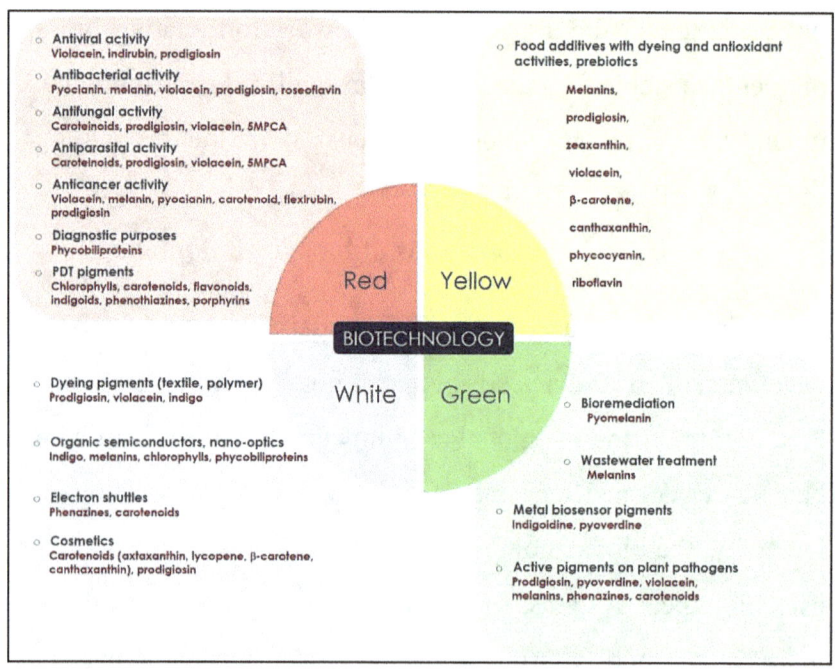

**Figure 3** Biotechnology rainbow cods representing the potential applications of bacterial pigments in each field (Venil et al., 2020).

### a. Bacterial pigments in red biotechnology

Red biotechnology focuses to find new drugs. In these times more than ever, the emergence of new viruses, microbial superbugs and multidrug-resistant infections and cancers, represent human life-threatening hazards. An investigation estimated that 10 million deaths due to antimicrobial resistance would be implemented every year after 2050. Additionally, in the same year, cancer incidence will double as a consequence of population growth and ageing. Furthermore, the current pandemic spread of SARS-CoV-2 is leading toward unpredictable and alarming scenarios. While biomedical research is principally aimed at obtaining an efficient vaccine, old and new drugs are administered to counteract the clinical manifestations of viral infection in different host tissues, comprising the respiratory, urinary, and cardiovascular systems. Among tentative drugs, bacterial pigments represent a promising reservoir to defeat pathogens as well as cancer (Orlandi et al., 2022; Agarwal et al., 2023). Figure (4) summarized the medical and pharmaceutical applications of bacterial pigments.

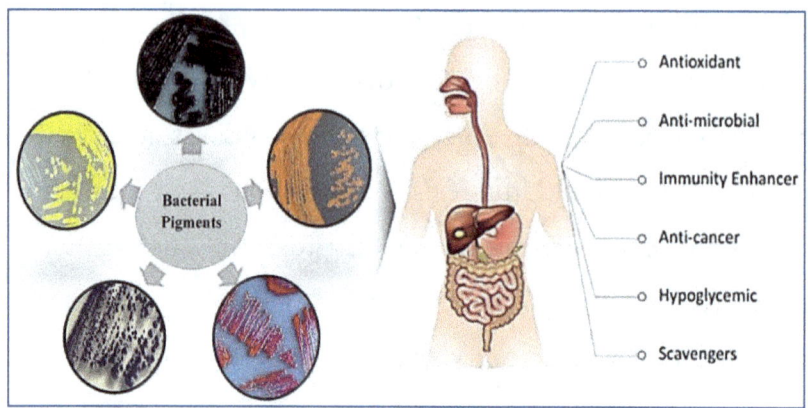

**Figure 4** Bacterial pigments from various genera and their proved bioactivities in red biotechnology (Sharma et al., 2021; Kumar et al., 2022).

i.  **Antibacterial pigments**

In recent years, infectious diseases had become the second major cause of death globally and third in developed countries after non-communicable disease. The increasing number of resistant microbes today had resulted in the high demand of newer antibiotics. Over the last 25 years, the number of new drugs entering the market had decreased and yet the prevalence of bacteria developing resistance continued to increase, causing difficulties to administer treatments. Continuous research to discover newer antimicrobial agents have been conducted and pigments are one good alternative. Bacterial pigments have been reported to have antimicrobial activities against Gram positive and Gram-negative bacteria. For example, *Virgibacillus halodenitrificans* producing carotenoids had been reported to demonstrate antimicrobial activities against antibiotic resistant *Staphylococcus aureus, Pseudomonas aeruginosa, Enterococcus faecalis* and *Escherichia coli* (Fayez et al., 2022). Besides, prodigiosin extracted from *Vibrio ruber* DSM 14379 can act as a bacteriostatic agent in treatment against Bacillus subtilis and *E. coli* (Choi et al., 2015). Moreover, prodigiosin produced by *S. marcescens* strongly inhibited the growth *of E. coli* and *S. aureus*, and caused severe leakage of intracellular substances in *S. aureus* (Matz et al., 2008). Whereas, pyocyanin, which is categorized among phenazines, was formerly defined as colicin owing to its ability to inhibit *E. coli* growth (Jayaseelan et al., 2014). In the same extend, violacein extracted from *C. violaceum* showed strong antibacterial activity against *S. aureus*, possibly via the disruption of the membrane integrity, as recently suggested (Aruldass et al., 2018).

Pyocyanin has been suggested to interfere with the cell membrane respiratory chain, thus impairing energy-requiring, membrane-bound metabolic processes, such as active transport into the cell. However, violaciens altered the permeability of the cytoplasmic membrane and, consequently, the physiology of the cell. Alteration of membrane integrity impairs the proton gradient and ATP synthesis; consequently, some essential functions, such as solute transport into the bacterial cell, DNA, and peptidoglycan synthesis, are compromised. Therefore, pigments behave more as disinfectants than antibiotics, as the former have a broader spectrum of activity than the latter against multidrug resistant strains (Orlandi et al., 2022). Similarly, Fayez et al., (2022) reported that the exact antimicrobial strategies for carotenoids are still ambiguous; however, the higher penetration through the cells suppresses microbial growth via possible scenarios including periplasmic membrane permeability, cytoplasmic content leakage, protein inhibition, arresting nucleic acids functionalities and generation of reactive oxygen species (ROS). It is worth mentioning that the combinations of light and pigments were found to be an effective strategy in antimicrobial approaches, as referred by (Chatragadda and Dufossé, 2021). In a study conducted by Leelanarathiwat et al., (2020), flavin mononucleotide activated by blue light resulted in inhibition of *S. aureus* biofilm. Via such strategy, i.e., photo-pigment therapy, the efficacy of microbial pigments against various pathogens could be easily increased.

### ii. Antifungal pigments

Fungal infections are the most common among immunocompromised, diabetic and immunodeficiency virus-infected populations. Unlike bacterial and viral pathogenesis, fungal pathogenesis is not complex, and fungal infections are comparatively easier to control. However, as the diseases prevalence has increased, more and more research has been performed to create potent new drugs to treat fungus infections. The main complexity in studying antifungal agents is that the drug should not interfere with the host's cellular mechanisms. Both the host and the fungal pathogen, being eukaryotes, make it difficult to trace and to target unique biomolecules, which will selectively harm the pathogen and not the host (Agarwal et al., 2023). Remarkably, pyocyanin produced on fermentation media displayed potent inhibition effect against the growth of *Candida albicans, C. parapsilosis* and *Cryptococcus* sp. (Sudhakar T, Karpagam, 2011). Likewise, a yellow-colored pigment produced by *Bacillus gibsonii* shows selective antifungal activity against economically essential fungi, and, with further investigations and clinical trials, it can be used as a prescription antifungal drug. It can also be formulated as an antifungal ointment (Dawoud et al., 2020). In the same context, *Neisseria spp.* create a distinctive red pigment that has excellent antifungal activity against a few key fungal species that are harmful to humans, including *Aspergillus spp., Candida spp.,* and *Trichoderma spp.* (Wagh et al., 2017). Furthermore, *P. aeruginosa* cultures produce extracellular pyocyanin, exhibiting potential antifungal activity without harming human consumption. As shown from the examples cited above, bacterial bioactive pigments possess huge potential to be used as fungicidal drugs if more research and trials are directed toward this area of bacterio-pharmacology.

### iii. Antiviral pigments

As new viruses are developing in humans consequently to spillover events, investigations on novel natural compounds with putative antiviral activity are compelling. Among them, violacein showed little inhibition of viral replication of Herpes Simplex Virus-1, Poliovirus type 2 and Simian rotavirus SA11 (Andrighetti-Fröhner et al., 2003). However, an indigoid pigment showed antiviral activity in human bronchial epithelial cells H292 infected with influenza virus A NWS/33 and B/Lee/40, by reducing both the expression and production of the chemokine RANTES (Mak et al., 2004). Prodigiosin was efficient in decreasing the viral titers of Bombyx mori nucleopolyhedrovirus (BmNPV), an enveloped double-stranded DNA virus with a high tropism for silkworms. PG was shown to behave as an inhibitor of DNA replication and transcription of BmNPV; mediated cell membrane fusion, which can block viral cell-to-cell transmission **(Zhou et al., 2016)**. Generally, as noticed from various literature, the antiviral results suggest that prodiginines may represent a new group of antiviral compounds.

### iv. Anticancer / Antineoplastic pigments

Toxicity, adverse events, and resistance represent major issues for oncology chemotherapy. Several reports in literature are available to demonstrate the possible efficacy of bacterial pigments in overcoming these limitations, by influencing apoptosis or autophagy pathways in cancer cell lines. Interestingly, species of the genus *Streptomyces* are known to produce many bioactive compounds and pigments that have demonstrated antitumour activity. In the search for a novel anticancer medication, *S. aburaviensis* and *S. psammoticus* are preferred possibilities (Ramirez-Rodriguez et al., 2018). *S. gresioaurianticus* JUACT 01 produces a yellow-colored pigment that induces apoptosis in HeLa and HepG2 cell lines and is nontoxic to the human lymphocytes (Prashanthi et al., 2015). Violacein is a naturally occurring bis-indole pigment with anti-tumour activity, which induces apoptosis in cancerous cell (Orlandi et al., 2022). On the other hand, prodigiosin pigments produced by *S. marcescens* have induced apoptosis in haematopoietic cancer cell lines and human colon cancer cells activities (Montaner et al., 2001). Significantly, violacein extracted from *C. violaceum* showed cytotoxic effects and apoptosis of different cancer cells including colorectal cancer, uveal melanoma, leukemia, and lymphoma cells in culture (Liu and Nizet, 2009). A yellow pigment producing *Pseudoalteromonas piscicida* strain NJ6-3-1 isolate obtained from sponge *Hymeniacidon perleve* possesses cytotoxic activity on cancer cells HeLa or BGC-823 cell lines, with IC50 values of 150 ±4.6 and 192 ±3.5 μg/mL, respectively (Zheng et al., 2006).

v. **Anti-inflammatory pigments**

Inflammation is a biological response that occurs in living tissues when they are subjected to injury. Anti-inflammatory drugs work by inhibiting the lysosomal enzyme that is released during inflammation. In this regard, several studies focused on collecting bacterial species from a marine environment. Through isolation, a diversity of pigments in shades of pink, yellow, orange, and brown were identified among the species. One such pigment, a bright yellow pigment, sourced from the species *Brevibacterium sp*, was chosen as the focal point for the study. The pigments' efficacy in reducing inflammation was evaluated using a model of induced paw edema in male Wister Albino rats. The results demonstrated a 100% effectiveness of the pigment in ameliorating inflammation (Gomes and Soares, 2023).

vi. **Antioxidant Activity**

Regardless of common carotenoids like lutein, β-carotene, astaxanthin, etc., the antioxidant activity of rare C50 carotenoids such as sarcinaxanthin, sarcinaxanthin mono glucoside, and sarcinaxanthin diglucoside with IC50 values of 57, 54, and 74 μM, respectively, were reported from a halophilic bacterium *Micrococcus yunnanensis* strain AOY-1 isolated from hard coral (Osawa et al., 2010). Violacein is a strong antioxidant compound that can protect lipid membranes from peroxidation caused by hydroxyl radicals (Stafsnes and Bruheim, 2013). Carotenoids with both large numbers of conjugated double bounds and hydroxyl groups appeared to have strong antioxidant activity. A structurally unusual phenolic carotenoid, 3,30-dihydroxyisorenieratene isolated from the bacterium *Streptomyces mediolani* (Martin et al., 2009), phycobiloproteins from cyanobacterial species have demonstrated powerful antioxidant activity. Cyanobacterial pigments such as β-carotene, lycopene, lutein C-phycocyanin, and phycobiliproteins are known to demonstrate antioxidant properties (Sonani et al., 2016).

### b. Bacterial pigments and white biotechnology

Different bacterial pigments find applications in industrial production as would be displayed.

### i. Solar cells and microbial fuel cells

Remarkably, the phototrophs possess molecular complexes with defined optical response regions that pave the way for photonic materials based on pigment assemblies. Some researchers suggested the use of bacterial chlorophylls in nanophotonics, for lasers. For example, a solar cell manufactured from this material would be able to function even on cloudy days. Similarly, the phycobiliproteins involved in bacterial photosystems light harvesting could be exploited in optics to assemble efficient light-trapping devices for capturing solar energy under low light. Carotenoid pigments isolated from the UV-resistant Antarctic (red) bacterium *Hymenobacter spp.* and (yellow) *Chryseobacterium spp.* belong to the xanthophyllin family and represent a promising tool as photosensitizers in the production of dye-sensitized solar cell technology (Orlandi et al., 2022). However, phenazines are a class of soluble pigments produced, among others, by *P. aeruginosa* and are characterized by electron transfer properties (Rabaey et al., 2005). This attitude seems to be exploited not only by producers, but also by other bacterial species, making these compounds a "collective good" to be a tool in biofuel cells, where electrons generated by microbial consortia are not directly transferred to cognate receptors but, instead, are diverted to an electrode with the production of electrical energy. Interestingly, constant oxidation of the electron shuttle phenazine causes overproduction of the dye concomitant with the enhancement of the bacterial concentration as presented by Orlandi et al., (2022).

ii. **Cosmetics**

Artificial or natural inorganic and organic pigments are primary sources of coloration in the cosmetics industry.

Nevertheless, bio-based colorants, in particular that of bacterial origin, are sought for sustainable concepts and additive functions such as antimicrobial and antioxidant activities in these bioproducts. Various bacterial genera and cyanobacteria are potent producers of high-value pigments such as carotenoids, phycocyanins, astaxanthin, lycopene, β-carotene, canthaxanthin and chlorophylls for developing product colors with tons of yellow, orange, red, green, and blue, which could be commercialized to some extent, thanks to their antioxidant properties. The application of bacteria for the production of pigments with other bioproducts, such as vitamins and lipids, has been supported by scientists due to their possible integration into biorefinery and sustainable production in cosmetic products including eyeliners, eye shadow powders, nail polish, and lipsticks (Kiki, 2023). Numerous genera of bacteria have been described and studied for melanin production, which can be applied in the cosmetics industry as an ingredient in dermal products, such as sunscreen, due to its high UV light protection properties. This property of protecting UV radiation in producing microbial cells can be extended to cosmetics when this pigment is extracted and included in a cosmetic formulation with photoprotective intent with the safe possibility of human use. The vast bioactive properties and different types of such biopigments demonstrate the potential of microbial pigments to be applied to the cosmetics industry, such as sunscreen, makeup, antiaging, skin lightening, and even for tattoos and permanent dyes combining the biological activity and coloring properties. In this regard, prodigiosin showed to increase by 20-65 % the sunscreen protection factor of dermatological creams. The addition of prodigiosin to extracts of Aloe vera leaf, and *Cucumis sativus* fruit increased of one order of magnitude the protection factor (Darshan N, Manonmani, 2015).

### iii. Textile dying

Applying bacterial pigments for dyeing fabrics is not a common practice in industries. It has still been little explored in scientific research. However, some more recent and initial work shows the use of these pigments in different fabric types in successive stages process of dipping in the pigment extract or boiling with the bacterial cells, color fixing, hot dyeing, washing and drying the dyed fabric (Barreto et al., 2023). However, the dyeing performances are different, depending on the types of fiber. From the colorfastness testing, the dyed fabrics also have the ability to maintain their color under several external conditions such as perspiration, washing, and rubbing/crocking. Color variation was achieved by changing the dipping time and the temperature of the dye bath. The color

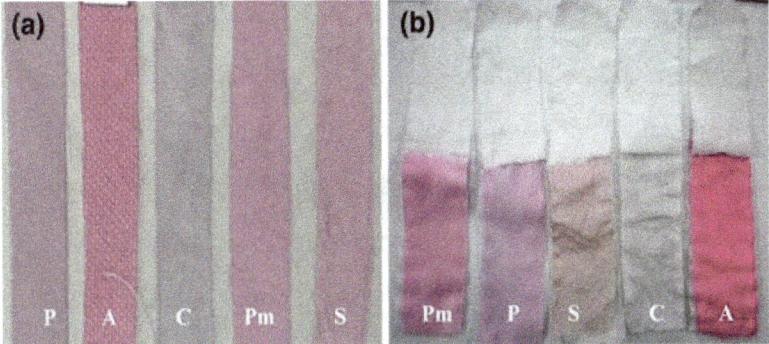

**Figure 5** Prodigiosin stained different fabrics, before (a) and after (b) washing; P polyester, A acrylic, C cotton, Pm polyester microfiber, S silk (Ahmad et al., 2012).

fastness of the dyed material was similar to vegetable pigment-dyed materials, however with lower light fastness (Ahmad et al., 2012). As illustrated in Ahmad et al., (2012), Prodigiosin from *Vibrio spp.* and *S. marcescens* was used as a textile colorant (Figure 5) for staining polyester, acrylic, cotton, polyester microfiber, and silk, the results showed that acrylic fiber dyed with excellent colorfastness toward washing with a rating of 5, followed by polyester microfiber (3/4), polyester (2/3), silk (2) and cotton (1). This is because acrylic fiber provided a better force of interaction with the dye and reduced its tendency to be washed out on laundering. Overall, a high colorfastness (staining) of 5 indicates that prodigiosin dyed on the fabrics would not stain other fabrics.

Similarly, textile-dyeing ability was also reported for *Janthinobacterium lividum* (Shirata et al. 2000), in which its purple pigment gave good color tone when applied on silk, cotton and wool (bluishpurple, all natural fibers), and nylon and vinylon (dark blue, both synthetic fibers). Let alone its antimicrobial activity against phytopathogenic fungi like *Rosellinia necatrix,* which causes white root rot of Mulberry. Additionally, violacein from *Pseudoalteromonas sp.* (DSM13623) was proposed for economical use in large amounts for consumer and environmental-friendly products, especially in textile and toy industries (Durán et al., 2012).

**iv. Candle and paper industry**

Commercial candles (fluted and translucent) were placed in a beaker and heated until completely melted before the addition of the bacterial culture broth. The mixtures were homogenized and poured into the mold. The wicks were immediately placed into the center of the mold and the candles were left to cool at room temperature for 1 h. Translucent candle showed a more intense coloration compared to the fluted candle as depicted in Figure (6).

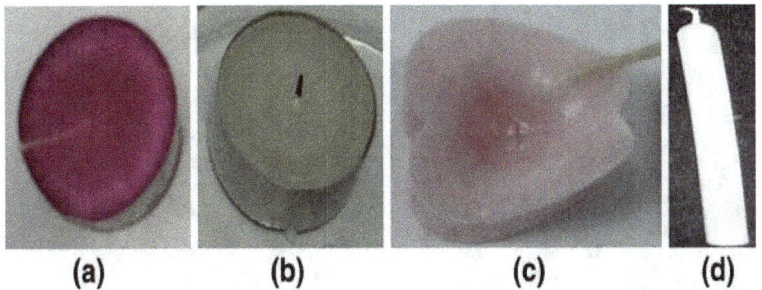

**Figure 6** Prodigiosin application in candle coloring (a) translucent candle and (c) fluted candle; (b) original translucent candle (d) original control fluted candle.

For assessing the potential application of prodigiosin in paper making, bacterial culture broth (3 mL) and one teaspoon of cornstarch were homogeneously mixed in one half of the thick pulp. However, the other half was not added with pigment (control). The pulp was then evenly spread onto a net to drain the excess water followed by 24 h drying. The prodigiosin – dyed paper was exposed to sunlight and fluorescent light for 4 h, whereas, paper unexposed to light acted as control. Initial intense red coloration on the paper was substantially reduced to faint red upon exposure to both sunlight and fluorescent light, with sunlight exerting the higher fading effect (Figure 7). This effect can be attributed to wider range of light wavelengths for sunlight compared to the fluorescent light (Ahmad et al., 2012).

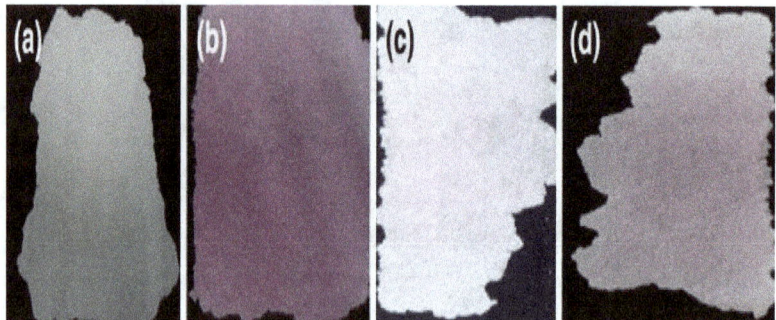

**Figure 7** Employment of prodigiosin in paper staining; (a) original paper (control), (b) paper dyed with prodigiosin, (c) dyed paper exposed to sunlight and (b) dyed paper exposed to fluorescent lamp (Ahmad et al., 2012).

4. **Conclusion and future perspectives**

In conclusion, bacterial pigments are found to be more valuable and demandable over synthetic compounds as potentially eco-friendly and biocompatibility with unique biological and industrial applications. Therefore, it is vital for bioprospecting about new producer strains from various and extreme niches. The more screened new species, the more the probability of finding unique pigments with enhanced activities. This will reduce our dependence on antibiotics and other chemotherapeutic agents and reduce patients' side effects due to their consumption. With the kind of lifestyle we are progressing into, bacterial pigments will protect us against diseases and reduce our dependency on synthetic and harmful coloring agents in food, medicine, and industry. However, the employment of synthetic biology and metabolic engineering strategies have been developed and optimized to increase the yield with reduced bioprocess costs. Different scenarios are recruited to build new biosynthetic pathways via recombining multiple target genes or regulating crucial molecule precursors in the biochemical metabolic chain to improve the final yield of a bio-pigment, which seems to be essential for the future market. The advancement of genetic engineering procedures, growth conditions, nutritional waste sources, and fermentation methods are grouped among such possible solutions. Intriguingly, multi-omic (genomics, transcriptomics, metabolomics, and proteomics) approaches with artificial intelligence were developed for better investigation of gene assembly of microbial agents in biotechnology. Another promising strategy of metabolic engineering is the employment of CRISPR/Cas technology. The pigment genes could be introduced into the bacterial gene using the CRISPR-Cas9 system to engineer and produce natural pigments. Noteworthy mention that the enhancement production of bacterial pigments should be accompanied by extensive toxicological studies (e.g., acute oral toxicity in mice 90-day subchronic toxicological study, acute dermal irritation, acute eye irritation), followed by assessing their antitumor activity, antibiotic activity, and ultimately obtaining regulatory approval (e.g., EU and USA legislations, Codex Alimentarius Commission, Food and Drug Administration, European Food Safety Authority, Pharmaceutical and Food Safety Bureau, and National Agency of Sanitary Vigilance). However, storage of pigments and their shelf life are suggested to be essential for commercialization to maintain high stability with maximum performance. Hereby, modifying encapsulation strategies should be under the circle of importance. In addition, the extraction techniques and solvents should be simplified and cost-effective. Ultimately, all efforts should be exerted simultaneously throughout the production process beginning from finding producer-strain passing through the production process and ending with commercialization; eventually, the consumer's opinions are highly required.

References

Agarwal H, Bajpai S, Mishra A, Kohli I, Varma A, Fouillaud M, Dufossé L, Joshi NC. Bacterial pigments and their multifaceted roles in contemporary biotechnology and pharmacological applications. Microorganisms. 2023, 28;11(3):614.

Ahmad WA, Wan Ahmad WY, Zakaria ZA, Yusof NZ, Ahmad WA, Ahmad WY, Zakaria ZA, Yusof NZ. Application of bacterial pigments as colorant. Springer Berlin Heidelberg; 2012.

Andrighetti-Fröhner CR, Antonio RV, Creczynski-Pasa TB, Barardi CR, Simões CM. Cytotoxicity and potential antiviral evaluation of violacein produced by Chromobacterium violaceum. Memórias do Instituto Oswaldo Cruz. 2003, 98:843-8.

Aruldass CA, Dufossé L, Ahmad WA. Current perspective of yellowish-orange pigments from microorganisms-a review. Journal of Cleaner Production. 2018 Apr 10;180:168-82.

Azman AS, Mawang CI, Abubakar S. Bacterial pigments: the bioactivities and as an alternative for therapeutic applications. Natural Product Communications. 2018 Dec;13(12):1934578X1801301240.

Barreto JV, Casanova LM, Junior AN, Reis-Mansur MC, Vermelho AB. Microbial Pigments: Major Groups and Industrial Applications. Microorganisms. 2023 Dec 4;11(12):2920.

Chatragadda R, Dufossé L. Ecological and biotechnological aspects of pigmented microbes: A way forward in development of food and pharmaceutical grade pigments. Microorganisms. 2021 Mar 18;9(3):637.

Choi SY, Yoon KH, Lee JI, Mitchell RJ. Violacein: properties and production of a versatile bacterial pigment. BioMed research international. 2015 Oct;2015.

Darshan N, Manonmani HK. Prodigiosin and its potential applications. Journal of food science and technology. 2015 Sep;52:5393-407.

Durán M, Ponezi AN, Faljoni-Alario A, Teixeira MF, Justo GZ, Durán N. Potential applications of violacein: a microbial pigment. Medicinal Chemistry Research. 2012 Jul;21:1524-32.

Fayez D, Youssif A, Sabry S, Ghozlan H, Eltarahony M. Carotegenic Virgibacillus halodenitrificans from Wadi El-Natrun Salt Lakes: isolation, optimization, characterization and biological activities of carotenoids. Biology. 2022 Sep 27;11(10):1407.

Gomes M, Soares G. A review of bacterial pigments: harnessing nature's colors for functional materials and dyeing processes. Textile & Leather Review. 2023; 6:167-190.

Hassan S., Abdrabo M. and elsayis A. Microbial pigments: sources and applications in the marine environment. Egyptian Journal of Aquatic Biology and Fisheries. 2022 Jan 1; 26(1):99-124.

Jayaseelan S, Ramaswamy D, Dharmaraj S. Pyocyanin: production, applications, challenges and new insights. World

Journal of Microbiology and Biotechnology. 2014 Apr;30:1159-68.

Kiki MJ. Biopigments of microbial origin and their application in the cosmetic industry. Cosmetics. 2023 Mar 10;10(2):47.

Kumar S, Kumar V, Nag D, Kumar V, Darnal S, Thakur V, Patial V, Singh D. Microbial pigments: learning from the Himalayan perspective to industrial applications. Journal of Industrial Microbiology and Biotechnology. 2022 Sep;49(5):kuac017.

Leelanarathiwat K, Katsuta Y, Katsuragi H, Watanabe F. Antibacterial activity of blue high-power light-emitting diode-activated flavin mononucleotide against Staphylococcus aureus biofilm on a sandblasted and etched surface. Photodiagnosis and Photodynamic Therapy. 2020 Sep 1;31:101855.

Liu GY, Nizet V. Color me bad: microbial pigments as virulence factors. Trends in microbiology. 2009 Sep 1;17(9):406-13.

Mak NK, Leung CY, Wei XY, Shen XL, Wong RN, Leung KN, Fung MC. Inhibition of RANTES expression by indirubin in influenza virus-infected human bronchial epithelial cells. Biochemical Pharmacology. 2004 Jan 1;67(1):167-74.

Martin HD, Kock S, Scherrer R, Lutter K, Wagener T, Hundsdörfer C, Frixel S, Schaper K, Ernst H, Schrader W, Görner H. 3, 3′-Dihydroxyisorenieratene, a natural carotenoid with superior antioxidant and photoprotective properties. Angew. Chem. 2009 Jan 2;48:400-3.

Matz C, Webb JS, Schupp PJ, Phang SY, Penesyan A, Egan S, Steinberg P, Kjelleberg S. Marine biofilm bacteria evade eukaryotic predation by targeted chemical defense. PloS one. 2008 Jul 23;3(7):e2744.

Montaner B, Pérez-Tomás R. Prodigiosin-induced apoptosis in human colon cancer cells. Life sciences. 2001 Mar 16;68(17):2025-36.

Narsing Rao MP, Xiao M, Li WJ. Fungal and bacterial pigments: secondary metabolites with wide applications. Frontiers in microbiology. 2017 Jun 22;8:250699.

Orlandi VT, Martegani E, Giaroni C, Baj A, Bolognese F. Bacterial pigments: A colorful palette reservoir for biotechnological applications. Biotechnology and Applied Biochemistry. 2022 Jun;69(3):981-1001.

Osawa A, Ishii Y, Sasamura N, Morita M, Kasai H, Maoka T, Shindo K. Characterization and antioxidative activities of rare C50 carotenoids-sarcinaxanthin, sarcinaxanthin monoglucoside, and sarcinaxanthin diglucoside-obtained from Micrococcus yunnanensis. Journal of Oleo Science. 2010;59(12):653-9.

Prashanthi K, Suryan S, Varalakshmi KN. In vitro anticancer property of yellow pigment from Streptomyces griseoaurantiacus JUACT 01. Brazilian Archives of Biology and Technology. 2015 Nov;58:869-76.

Rabaey K, Boon N, Höfte M, Verstraete W. Microbial phenazine production enhances electron transfer in biofuel cells. Environmental science & technology. 2005 May 1;39(9):3401-8.

Rajendran P, Somasundaram P, Dufossé L. Microbial pigments: Eco-friendly extraction techniques and some industrial applications. Journal of Molecular Structure. 2023 Oct 15;1290:135958.

Ramesh C, Prasastha VR, Venkatachalam M, Dufossé L. Natural substrates and culture conditions to produce pigments from potential microbes in submerged fermentation. Fermentation. 2022 Sep 14;8(9):460.

Ramirez-Rodriguez L, Stepanian-Martinez B, Morales-Gonzalez M, Diaz L. Optimization of the cytotoxic activity of three Streptomyces strains isolated from guaviare river sediments (Colombia, South America). BioMed Research International. 2018 Jul 19;2018.

Sánchez-Muñoz S, Mariano-Silva G, Leite MO, Mura FB, Verma ML, da Silva SS, Chandel AK. Production of fungal and bacterial pigments and their applications. InBiotechnological production of bioactive compounds 2020 Jan 1 (pp. 327-361). Elsevier.

Sharma M, Usmani Z, Gupta VK, Bhat R. Valorization of fruits and vegetable wastes and by-products to produce natural pigments. Critical Reviews in Biotechnology. 2021 May 19;41(4):535-63.

Shirata A, Tsukamoto T, Yasui H, Hata T, Hayasaka S, Kojima A, Kato H. Isolation of bacteria producing bluish-purple pigment and use for dyeing. Japan Agric. Res. Quart. 2000, 34:131–140.

Slama HB, Chenari Bouket A, Pourhassan Z, Alenezi FN, Silini A, Cherif-Silini H, Oszako T, Luptakova L, Golińska P, Belbahri L. Diversity of synthetic dyes from textile industries, discharge impacts and treatment methods. Applied Sciences. 2021, 6;11(14):6255).

Sonani RR, Rastogi RP, Patel R, Madamwar D. Recent advances in production, purification and applications of phycobiliproteins. World journal of biological chemistry. 2016 Feb 2;7(1):100.

Stafsnes, M.H.; Bruheim, P. Pigmented Marine Heterotrophic Bacteria. In Marine Biomaterials: Characterization, Isolation and Applications; Kim, S., Ed.; CRC Press, Taylor & Francis Group: London, UK, 2013; pp. 117–148.

Sudhakar T, Karpagam S. Antifungal efficacy of pyocyanin produced from bioindicators of nosocomial hazards. InInternational Conference on Green Technology and

Environmental Conservation (GTEC-2011) 2011 Dec 15 (pp. 224-229). IEEE.

Usman HM, Abdulkadir N, Gani M, Maiturare HM. Bacterial pigments and its significance. MOJ Bioequivalence & Bioavailability. 2017;4(3):00073.

Venil CK, Dufossé L, Renuka Devi P. Bacterial pigments: sustainable compounds with market potential for pharma and food industry. Frontiers in Sustainable Food Systems. 2020 Jul 21;4:100.

Venil CK, Zakaria ZA, Ahmad WA. Bacterial pigments and their applications. Process Biochemistry. 2013 Jul 1;48(7):1065-79.

Vinotha M, Prabhavathi P, Vijayaraghavan R, Kumar D. Bacterial pigments and their application in textile industries using mordants. Journal of Advanced Scientific Research. 2019 Sep 10;10(03 Suppl 1):139-45.

Wagh P, Mane R. Identification and characterization of extracellular red pigment producing Neisseria spp. isolated from soil sample. Int. J. Innov. Knowl. Concept. 2017;5:23-5.

Zheng L, Yan X, Han X, Chen H, Lin W, Lee FS, Wang X. Identification of norharman as the cytotoxic compound produced by the sponge (Hymeniacidon perleve)-associated marine bacterium Pseudoalteromonas piscicida and its apoptotic effect on cancer cells. Biotechnology and applied biochemistry. 2006 Jun;44(3):135-42.

Zhou W, Zeng C, Liu R, Chen J, Li R, Wang X, Bai W, Liu X, Xiang T, Zhang L, Wan Y. Antiviral activity and specific modes of action of bacterial prodigiosin against Bombyx mori nucleopolyhedrovirus in vitro. Applied microbiology and biotechnology. 2016 May;100(9):3979-88.

# CHAPTER II

# Biopigments From Actinomycetes and Related Microbes: Applications and Diverse Activities

**Tarek Hosny Taha,** *PhD*

Environmental Biotechnology Department, Genetic Engineering and Biotechnology Research Institute (GEBRI), City of Scientific Research and Technological Applications, New Borg El-Arab City, 21934, Alexandria, Egypt.

## Abstract

Biopigments are natural products produced by multiple organisms including microorganisms such as actinomycetes and plants. These products have recently gained a lot of attention as safe, low-cost, eco-friendly alternative substances that can replace the currently used synthetic ones. They have a lot of applications and advantages that help for their intensive incorporation in the industrial applications such as: food colorants and cosmetics. In addition, they also have a lot of medical and pharmaceutical application as they have antimicrobial, antioxidant, anti-inflammatory, and anti-cancer activities.

**Keywords:** Biopigments, Actinomycetes, Applications of biopigments, Activities of biopigments.

## Introduction

The word 'actinomycetes' is basically extracted from the Greek words "atiks" which means 'a ray' and "mykes" which means 'fungus' (Das et al. 2008). According to this definition they are group of microorganisms that gain the characteristics of both fungi and bacteria. At the beginning of their discovery they were classified as an intermediate group between the bacteria and the fungi, however, after some time they have been separated as a different group (Pandey et al. 2004). They are present in both aquatic and terrestrial habitats and characterized by the formation of thread-like filaments when growing in soil, in addition to forming hyphae and spores like fungi and being Gram positive bacteria with high content of G and C nucleotides (Das et al. 2008; Pandey et al. 2004; Berdy 2005; Lam 2006).

They can potentially produce various economically valuable metabolites such as enzymes, antibiotics, anti-tumor agents, and many other beneficial products (Figure 1) (Berdy 2005; Lam 2006; Sarkar and Suthindhiran 2022).

Actinomycetes are belonging to the class Actinobacteria of phylum Firmicutes, that is divided into eight families (Actinomycetaceae, Actinoplanaceae, Dermatophilaceae, Mycobacteriaceae, Frankiaceae, Nocardiaceae, Micromonosporaceae, and Streptomyceteceae) with more than 60 genera (Hameş-Kocabaş and Uzel 2007). They have rigid cell wall with a composition similar to Gram positive bacteria which consists of peptidoglycan, polysaccharides, in addition to the teichoic and teichuronic acid (Mahon and Lehman 2022; Goodfellow et al. 1998; Das et al. 2008).

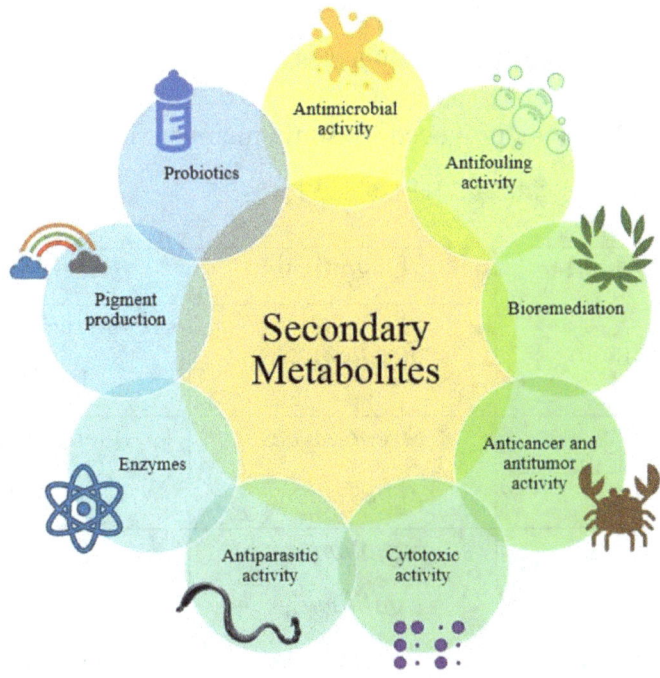

**Figure 1** Beneficial secondary metabolites produced by marine actinomycetes (Sarkar and Suthindhiran 2022).

Actinomycetes are able to produce various types of secondary metabolites with multiple biological activities (Berdy 2005). For instance, the genus Streptomyces can produce more than 10,000 metabolites including volatile organic compounds (van Wezel et al. 2009; Hopwood 2007). In addition to the metabolites with clinical significance, they have also reported to produce industrially important metabolites such as color pigments. These naturally produce pigments are currently used for different purposes of the food and pharmaceutical industries (Prakash et al. 2001). Table 1 is representing some of the pigments produced by different members of actinomycetes.

**Table 1** Pigments produced by some members of actinomycetes (Rinkal and Srivathsa 2018).

| Pigment | Organism | Reference |
|---|---|---|
| Melanin | *Streptomyces* | (Conn and Conn 1941) |
| Naphthoquinone | *Streptomyces coelicolor* | (Gerber and Wieclawek 1966) |
| Anthracyclin glycoside | *Streptomyces galilaeus, Streptomyces melanogenes, Streptomyces peucetius* | (Cassinelli et al. 1982) |
| Phenoxazinone | *Streptomyces parvullus* | (Smith et al. 2004) |

| | | |
|---|---|---|
| Carotenoids | *treptomyces griseus, Streptomyces setonii, Streptomyces coelicolor* | (Takano et al. 2006) |
| Violacein | *Chromobacterium violaceum* | (Durán et al. 2007) |
| Prodigiosin | *Serratia, S treptoverticillium rubrireticuli, Streptomyces longisporus* | (Chidambaram and Perumalsamy 2009) |

**Production of natural pigments by actinomycetes**

In most cases, natural pigments are produced by microorganisms under certain environmental conditions. A lot of microbial species producing pigments when their colonies become age or when certain components are available in the culture media. The produced pigments are either soluble in water or soluble in oil. The water-soluble pigments are spontaneously spread through the culture media in which they already produced. However, in case of oil soluble pigments, the pigmented colony is required to be shaken in oil, where the oil will be pigmented if the pigments are fat soluble (BAWAZIR and PALAKSHA 2019).

*Streptomyces coelicolor* and *Streptomyces violaceoruber* are soil actinomycetes that produce the blue pigments that are extensively applied in the industrial, scientific, and medical sectors (Palanichamy et al. 2011). Multiple pigments with multiple colors such as carotenoids, flavins, indigo, monacens, and melanins have been reported by different microorganisms. Among these pigments, carotenoids are produced by different bacteria, fungi, and plants with a specific yellow-to-orange red color (Goodwin and Britton 1988). They have intensively and commercially applied in recent days in the fields of beauty care, nutraceuticals, and other pharmaceutical purposes (Klein-Marcuschamer et al. 2007). They have reported as cancer suppressors and can enhance the immunity responses, as well (Krinsky and Johnson 2005). They have propitamin action and antioxidant activities that help to prevent the life-related diseases (Young and Lowe 2001).

On the other hand, the pigments produced by bacteria have different chemical structures and a lot of colors that shade the color of their producing colonies. For instance, *Serratia marcescens* is represented by red color, *Chryseoacterium artocarpi* is represented by red-yellow color, *Pseudomonas* sp. is represented by green color (Venil et al. 2013). In general, most of the common pigments and their colors are represented in Figure 2.

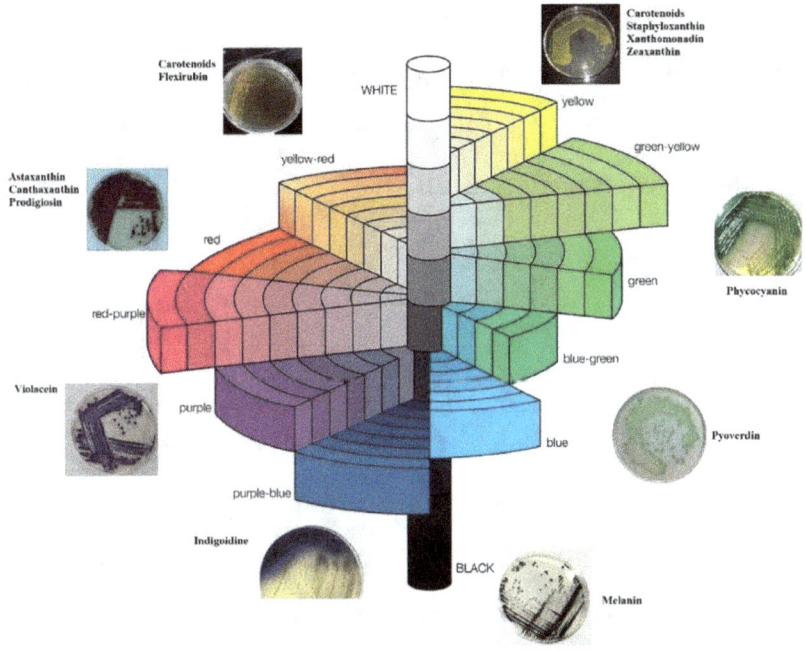

**Figure 2** The common microbial pigments and their colors (Venil et al. 2013).

**Industrial applications of pigments**

- **As food colorants**

Marine microorganisms can commercially produce high food-grade pigments with little or no threats to the general health of the consumers. They are representing pleasant colors even when used in low concentrations such as pyocyanin and pyorubrin that are produced by *P. aeruginoasa* sp., and giving pleasant colors at 25 mg/ml/g agar (Saha et al. 2008). They have been also used as feed additives in order to activate the growth and improve the coloration of ornamental marine creatures (Dharmaraj et al. 2009). Moreover, some pigments such as prodigiosin extracted from marine microorganism and showed acceptable staining properties and long shelf life (Ramesh et al. 2020).

- **As dying agents**

Recent cloth designs are requesting fabrics with antimicrobial properties. Using of bio-pigments of the microorganisms can provide dual beneficial properties as coloring agents and as antimicrobial agents. From this point of view, Lee and his colleagues succeeded to produce prodigiosin and cycloprodigiosin pigments from the marine bacteria *Z. rubidus* sp. S1-1 that showed antimicrobial activities against *S. aureus* sp. KCTC 1916 and *E. coli* sp. KCTC 1924 beside their ability to stain the cotton and silk fabrics (Lee et al. 2011). Similarly, the marine *Vibrio* sp. produced the prodiginine pigment that is appeared as bright red color and able to stain some fabrics such as silk, nylon 66, acrylic and wool in addition to stop the microbial growth of *S. aureus* and *E. coli* (Alihosseini et al. 2008). In addition, *C. violacea* microbial strain can also produce violacein pigment that added to fabrics and enhance their antibacterial properties (Nawaz et al. 2020). It has been reported that prodigiosin dye produced by *Serratia* sp. BTWJ8 was able to stain papers, rubber, and some polymer sheets such as PMMA (Polymethyl methacrylate) (Krishna et al. 2013).

- **As cosmetics ingredient**

As a global emerging business market, cosmetics industry has a worth of tens of billions in France, Germany, and UK especially the skincare products (Camillo et al. 2018). Multiple researches have believed that the pigments produced by marine microorganisms can play an important role in cosmetic industry including skincare products. Some of these pigments can protect against the UV rays and can increase the sun protection factor (SPF) to 30% (Patil et al. 2016). For instance, F3 cream is containing melanin pigment synthesized by the bacteria *Halomonas venusta* sp. and the seaweed *Gelidium spinosum* and showed photoprotective activity and elevated SPF in addition to its efficiency in wound healing and antibacterial activity against *Streptococcus pyogenes* sp. and *S. aureus* sp. [127]. Some of these pigments have been also reported as protectors for the mammalian cells from UV irradiation. *Vibrio natriegens* sp.is a marine bacterial species that was able to produce melanin pigment that helped 90% of HeLa cells to survive in melanized cell culture (Wang et al. 2019).

- **As antifouling agent**

The fouling activities of marine vehicles is a serious problem that causing the spent of billions of dollars each year. This biofouling is increasing the roughness and promoting the frictional resistance which leads to increasing the fuel consumption and other related environmental compliances (Nawaz et al. 2020). Some of the commercially used antifoulants are containing heavy metals and causing severe environmental problems. Recent studies reported the using of marine microbial pigments as eco-friendly antifouling agents. In this regard, *Serratia* sp. has been reported to produce prodigiosin that can play an important role as antifouling agent against *Gallionella* sp. and *Alteromonas* sp. which are known as potent fouling bacterial species. In addition, it is able to stop the ability of *Cyanobacterium* sp. to adhere the glass surfaces (Priya et al. 2013). Polymelanin is another example of the pigments produced by the marine bacteria *P. lipolytica* sp. that able to prevent the metamorphosis and the invertebrate larval settlement (Zeng et al. 2017). In general, the bioproduction scenario of biopigments by microorganisms and their applications can be summarized in Figure 3.

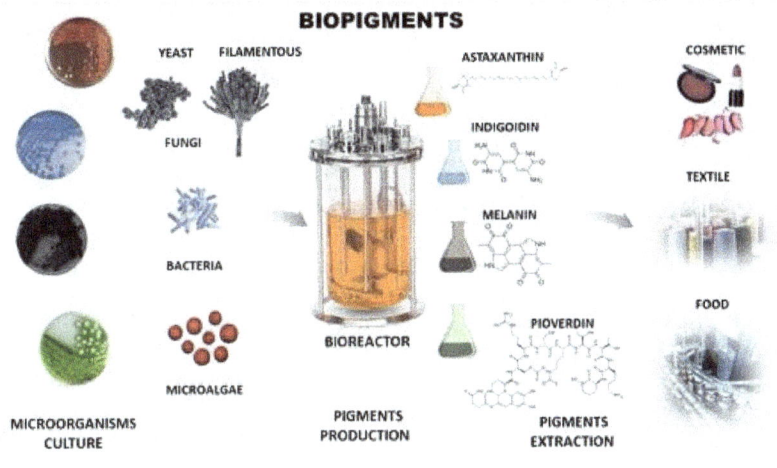

**Figure 3** General scenario of the production and applications of pigments produced by microbes (Barreto et al. 2023).

**Types of biopigments activities**

- **Pharmacological activity**

Biopigments have been reported as anticancer, antimicrobial, and immune-suppressive agents, in addition to their ability to treat various diseases. They have been also used for the diagnosis of many diseases such as leukemia, cancer, and diabetes mellitus (Kumar et al. 2015). As mentioned before, some of the microbial pigments such as melanin have been used for the protection of human skin from UV radiation, which makes it an essential component of the sun cream blocks (Narsing Rao et al. 2017). Some of the common microbial pigments such as pyocyanin (blue-green), violacein (violet), prodigiosin (red), and carotenoids (yellow - orange) have showed different antioxidant, anticancer, antiviral, antitumor, and antibacterial activities.

It has been reported that the highest antimicrobial activities were detected for red pigments followed by orange, yellow, and green pigments (Azamjon et al. 2011). The red pigment (Astaxanthin) is belonging to carotenoids and has high contribution in the fields of pharmaceuticals and feed industry. Similarly, the xanthophylls can be used as nutraceutical agent that able to prevent cancers, heart attacks, and strokes (Long 2004; Kim et al. 2012). Prodigiosin has also been reported as antifungal, antiprotozoal, antibacterial agent, in addition to its anti-inflammatory, cytotoxic, and anticancer properties (Li et al. 2018; Panesar et al. 2015).

- **Antimicrobial activity**

The human infections caused by the bacteria resistant to commercial antibiotics are serious human life-threatening agents (Van Duin and Paterson 2016). To overcome this issue, the discovery of new antimicrobial agent is necessary. Potent antimicrobial properties have been reported by microbial pigments such as melanins, flavins, and carotenoids (Venil et al. 2013). The orange-pigmented bacteria (*Pseudoalteromonas flavipulchra*) has showed potent antimicrobial activity against some of the drug resistant microbes such as *E. coli*, MRSA, *Acinetobacter baumannii*, and *Enterobacter aerogenes* (Ayuningrum et al. 2017). Some of the potent antimicrobial agents such as marino-quinolines A, C and D, marinoquinoline I, marinopyrazinone B, and marinoazepinone B have been produced by marine pigments microorganism (Choi et al. 2015; Romanenko et al. 2015; Kalinovskaya et al. 2017).

β-carotene (yellow pigment) produced by *Vibrio owensii* was reported as antimicrobial agent against *Klebsiella pneumoniae*, *E. coli*, and MRSA pathogens (Sibero et al. 2019). A pinkish-orange pigment produced by the marine strain *Salinicoccus* sp. showed high antimicrobial activity against *Staphylococcus aureus* (Srilekha et al. 2017). Similarly, the prodigiosin pigment extracted from *Chromobacterium prodgiosum* showed promising results against bacteria such as *Staphylococcus aureus* and *Corynebacterium diphtheriae* (Mumtaz et al. 2019), while the same pigment extracted from *Neisseria* sp. showed promising results against some fungi such as *Aspergillus* sp., and *Trichoderma* sp. (Wagh and Mane 2017). In addition, *Halomonas* sp. was able to produce carotenoids pigments with antimicrobial activities against *Streptococcus pyogenes*, *P. aeruginosa*, and *Klebsiella* sp. (Ravikumar et al. 2016). Moreover, violacein and deoxyviolacein has been reported as antimicrobial agents against the pathogenic fungus *Rosellinia necatrix* (Bisht et al. 2020).

- **Anticancer activity**

The sever cancer diseases are still depending on chemotherapy for their treatment. Chemotherapy is badly affecting the normal cells as well, which worsens the patient's recovery. The finding of cheaper and safer natural products in order to treat cancer disease is urgently demanded (Parmar et al. 2015; Sharma et al. 2017). Recent studies reported that natural microbial pigments have huge potential as anticancer compounds and hence their anticancer trials deserve comprehensive investigations (Srilekha et al. 2018). The natural microbial pigment prodigiosin succeeded to induce apoptosis in different cancer cell lines such as HSC-2 cells (human oral cancer cells) and hematopoietic cell lines (Jurkat, NSO and HL60). Similarly, prodigiosin extracted from *Pseudoalteromonas* sp. 1,020 showed cytotoxic effect against U937 leukemia cells (Wang et al. 2012).

In another study, the yellowish pigment produced by *Rhodococcus maris* showed a reduction in the breast cancer spreading (Elsayed et al. 2017). Also, the red pigment extracted from *Athrobacter* sp. G20 investigated anticancer activity against esophageal cancer cell lines (Afra et al. 2017), in addition to the anticancer activity against breast cancer cell lines MCF-7 which recorded through the using of carotenoids from *Kocuria* sp. QWT-12 (Rezaeeyan et al. 2017). Violacein has also reported anticancer activities against multiple cell lines such as V79 fibroblasts (da

Silva Melo et al. 2000) and human breast cancer cells (Alshatwi et al. 2016), in addition to its induced activation of the inflammatory responses (Venegas et al. 2019). In another study, Silva and his team reported that the cancer cells that known for their resistance to chemotherapy can be killed by the phycocyanin pigment which able to interact with non-specific targets from membranes to nuclei (e Silva et al. 2018). Moreover, melanin pigment produced by *Streptomyces glaucescens* NEAE-H was reported as an anticancer agent against HFB4 skin cancer cell line (El-Naggar and El-Ewasy 2017).

- **Antioxidant, Anti-allergic and anti-inflammatory activities**

The antioxidants are known for their free radicals scavenging capacity which help for the protection of human against many infections and degenerative diseases. The currently common antioxidants are synthetic ones which have potential health hazards, and push for figuring out natural alternatives (Lee et al. 2014). Some of the microbial pigments such as carotenoids produced by *Kocuria marina* and *Thermus* strains have reported potent antioxidant activities (Rezaeeyan et al. 2017). It has been also reported that the carotenoids extracted from *Pedobacter* and *Fontibacter flavus* YUAB-SR-25 showed potent antioxidant capacity with simultaneous inhibition of lipid peroxidation (Correa-Llantén et al. 2012; Prabhu et al. 2013).

Moreover, β-carotene has suppressed the bad effects of the free radicals in human (Fiedor and Burda 2014). Furthermore, the carotenoids extracted from *Planococcus* sp. TRC1 showed significant antioxidant activity that lead to its industrial applicability (Majumdar et al. 2019). On the other hand, Srilenkha and his team reported the isolation of Micrococcus sp. which is marine pigmented bacteria that recorded strong anti-inflammatory effect and wound healing capacity (Srilekha et al. 2018). Similarly, Egeland reported the extraction of a bioactive carotenoid named fucoxanthin from cyanobacteria with potent anti-inflammatory, wound healing, and anti-obesity properties (Rossi and De Philippis 2016).

## Stability of biopigments

However, biopigments have a lot of applications, they are facing limited marketing as a result of their poor stability. Multiple techniques have been evolved in order to increase the stability and shelf life of these natural colorants in a cost-effective manner (Sen et al. 2019). The most common and successful techniques used to address the stability issues are encapsulation and nano-emulsion.

- ### Encapsulation

The microbial pigments are known as unstable compounds when they exposed to certain conditions such as high temperatures, oxygen, and light which make them lose some of their characteristics (Narsing Rao et al. 2017). Increasing the solubility and stability of pigments can be achieved through the microencapsulation method as an alternative effective one that entraps these active components inside the microparticles. This method can entrap the compounds in its solid, liquid, or gas phase in the microparticles that have size range from nm to mm. When the pigments entrapped inside the packaging material it will be the core, while the packaging material will be the outer shell. This shell should have some important properties such as biodegradable, has low hygroscopicity, has low viscosity, and has emulsifying properties (Barros and Stringheta 2006), in addition to being include modified starch or other similar substances (Özkan and Bilek 2014).

After encapsulation, the pigments almost showed better and improved stability which increasing their shelf life, as the outer shell substances would protect the core substance from surrounding conditions such as humidity, oxygen, temperature, and light in addition to controlling the release of the pigments and preventing the leakage of their aroma. Some pigments such as anthocyanin has been encapsulated by maltodextrin using spray drying (Silva et al. 2013), and β-carotene which encapsulated in modified starch using freeze drying (Spada et al. 2012). Some reports showed better stability of the microencapsulation of flexirubin extracted from *Chryseobacterium artocarpi* CECT8497 compared with the free one. Moreover, Namazkar and his team reported an increased stability of encapsulated prodigosin from *Serratia marcescen* against pH, light, and temperature compared with the nonencapsulated form (Namazkar et al. 2013).

- **Nano-emulsion**

Nano-emulsion is another technique that is used to encapsulate the microbial pigments. Nano-emulsions are droplet of the size of 100 nm or less, and includes water, oil, and emulsifier. The presence of the emulsifier is crucial as it decreases the tension between the water and the oil and hence affect the formation of the nano-emulsion droplets. Surfactants are the most used emulsifiers; however, lipids and proteins can also be used. Nano-emulsions are more preferred than micro or macro-emulsions; as they have large surface area and more resistant to physical and chemical changes (Gupta et al. 2016), in addition to being non-irritant and nontoxic when used in food applications (Jaiswal et al. 2015). The stability of β-carotene pigment has been increased when entrapped in nano-emulsion using β-lactoglobulin as an emulsifier (Yi et al. 2014).

## Conclusion

The microbial biopigments has recently gained a lot of attention due to their multiple advantages and safety for the environment and for the human health. They have been industrially applied as food colorants, dying agent, ani-fouling agent, and as a cosmetic ingredient. They have showed antimicrobial, ani-cancer, anti-allergic, and anti-inflammatory activities that assist their participation in the medical and pharmaceutical markets. Increasing their stability and shelf life have been improved through their encapsulation in polymeric matrices or in nano-emulsions.

## References

Afra S, Makhdoumi A, Matin M, Feizy J (2017) A novel red pigment from marine Arthrobacter sp. G20 with specific anticancer activity. Journal of applied microbiology 123 (5):1228-1236

Alihosseini F, Ju KS, Lango J, Hammock BD, Sun G (2008) Antibacterial colorants: characterization of prodiginines and their applications on textile materials. Biotechnology progress 24 (3):742-747

Alshatwi AA, Subash-Babu P, Antonisamy P (2016) Violacein induces apoptosis in human breast cancer cells through up regulation of BAX, p53 and down regulation of MDM2. Experimental and Toxicologic Pathology 68 (1):89-97

Ayuningrum D, Kristiana R, Asagabaldan M, Sabdono A, Radjasa O, Nuryadi H, Trianto A Isolation, characterisation and antagonistic activity of bacteria symbionts hardcoral Pavona sp. isolated from Panjang Island, Jepara against infectious multi-drug resistant (MDR) bacteria. In: IOP Conference Series: Earth and Environmental Science, 2017. vol 1. IOP Publishing, p 012029

Azamjon B, Hosokawa S, Enomoto K (2011) Bioactive pigments from marine bacteria: applications and physiological roles. Evid Based Complementary Alternative Med 17

Barreto JVdO, Casanova LM, Junior AN, Reis-Mansur MCPP, Vermelho AB (2023) Microbial Pigments: Major Groups and Industrial Applications. Microorganisms 11 (12):2920

Barros F, Stringheta PC (2006) Microencapsulamento de antocianinas: Uma alternativa para o aumento de sua aplicabilidade como ingrediente alimentício. Biotecnologia ciência e desenvolvimento 36 (36):18-24

BAWAZIR AMA, PALAKSHA MS (2019) An Implication Of Actinomycetes On Human Well-Being: A Review. Int J Pharm Pharm Sci

Berdy J (2005) Bioactive microbial metabolites. The Journal of antibiotics 58 (1):1-26

Bisht G, Srivastava S, Kulshreshtha R, Sourirajan A, Baumler DJ, Dev K (2020) Applications of red pigments from psychrophilic Rhodonellum psychrophilum GL8 in health, food and antimicrobial finishes on textiles. Process biochemistry 94:15-29

Camillo F, Rota A, Biagini L, Tesi M, Fanelli D, Panzani D (2018) The current situation and trend of donkey industry in Europe. Journal of Equine Veterinary Science 65:44-49

Cassinelli G, RIVOLA G, Ruggieri D, ARCAMONE F, GREIN A, MERLI S, SPALLA C, CASAZZA AM, DI MARCO A, Pratesi G (1982) New anthracycline glycosides: 4-O-demethyl-11-deoxydoxorubicin and analogues from Streptomyces peucetius var. aureus. The Journal of Antibiotics 35 (2):176-183

Chidambaram K, Perumalsamy L (2009) An Insightful overview on microbial pigment, prodigiosin. Electronic journal of biology 5 (3):49-61

Choi EJ, Nam S-J, Paul L, Beatty D, Kauffman CA, Jensen PR, Fenical W (2015) Previously uncultured marine bacteria linked to novel alkaloid production. Chemistry & biology 22 (9):1270-1279

Conn H, Conn JE (1941) Value of pigmentation in classifying Actinomycetes: a preliminary note. Journal of Bacteriology 42 (6):791-799

Correa-Llantén DN, Amenábar MJ, Blamey JM (2012) Antioxidant capacity of novel pigments from an Antarctic bacterium. Journal of Microbiology 50 (3):374-379

da Silva Melo P, Maria SS, Vidal BdC, Haun M, Durán N (2000) Violacein cytotoxicity and induction of apoptosis in V79 cells. In Vitro Cellular & Developmental Biology-Animal 36 (8):539-543

Das S, Lyla P, Ajmal Khan S (2008) Distribution and generic composition of culturable marine actinomycetes from the sediments of Indian continental slope of Bay of Bengal. Chinese Journal of Oceanology and Limnology 26:166-177

Dharmaraj S, Ashokkumar B, Dhevendaran K (2009) Food-grade pigments from Streptomyces sp. isolated from the marine sponge Callyspongia diffusa. Food Research International 42 (4):487-492

Durán N, Justo GZ, Ferreira CV, Melo PS, Cordi L, Martins D (2007) Violacein: properties and biological activities. Biotechnology and applied biochemistry 48 (3):127-133

e Silva EF, Figueira F, Cañedo A, Machado K, Salgado M, Silva T, Wagner E, Mattozo F, Lima É, Sales-Neto J (2018) C-phycocyanin to overcome the multidrug resistance phenotype in human erythroleukemias with or without interaction with ABC transporters. Biomedicine & Pharmacotherapy 106:532-542

El-Naggar NE-A, El-Ewasy SM (2017) Bioproduction, characterization, anticancer and antioxidant activities of extracellular melanin pigment produced by newly isolated microbial cell factories Streptomyces glaucescens NEAE-H. Scientific reports 7 (1):42129

Elsayed Y, Refaat J, Abdelmohsen UR, Fouad MA (2017) The Genus Rhodococcus as a source of novel bioactive substances: A review. Journal of Pharmacognosy and Phytochemistry 6 (3):83-92

Fiedor J, Burda K (2014) Potential role of carotenoids as antioxidants in human health and disease. Nutrients 6 (2):466-488

Gerber NN, Wieclawek B (1966) The Structures of Two Naphthoquinone Pigments from an Actinomycete1. The Journal of Organic Chemistry 31 (5):1496-1498

Goodfellow M, Stainsby FM, Davenport R, Chun J, Curtis T (1998) Activated sludge foaming: the true extent of actinomycete diversity. Water Science and Technology 37 (4-5):511-519

Goodwin TW, Britton G (1988) Distribution and analysis of carotenoids. Plant pigments:61-132

Gupta A, Eral HB, Hatton TA, Doyle PS (2016) Nanoemulsions: formation, properties and applications. Soft matter 12 (11):2826-2841

Hameş-Kocabaş EE, Uzel A (2007) Alkaline protease production by an actinomycete MA1-1 isolated from marine sediments. Annals of microbiology 57:71-75

Hopwood DA (2007) Streptomyces in nature and medicine: the antibiotic makers. Oxford University Press,

Jaiswal M, Dudhe R, Sharma P (2015) Nanoemulsion: an advanced mode of drug delivery system. 3 Biotech 5:123-127

Kalinovskaya NI, Romanenko LA, Kalinovsky AI (2017) Antibacterial low-molecular-weight compounds produced by the marine bacterium Rheinheimera japonica KMM 9513 T. Antonie Van Leeuwenhoek 110:719-726

Kim HW, Kim JB, Cho SM, Chung MN, Lee YM, Chu SM, Che JH, Kim SN, Kim SY, Cho YS (2012) Anthocyanin changes in the Korean purple-fleshed sweet potato, Shinzami, as affected by steaming and baking. Food chemistry 130 (4):966-972

Klein-Marcuschamer D, Ajikumar PK, Stephanopoulos G (2007) Engineering microbial cell factories for biosynthesis of isoprenoid molecules: beyond lycopene. TRENDS in Biotechnology 25 (9):417-424

Krinsky NI, Johnson EJ (2005) Carotenoid actions and their relation to health and disease. Molecular aspects of medicine 26 (6):459-516

Krishna JG, Jacob A, Kurian P, Elyas K, Chandrasekaran M (2013) Marine bacterial prodigiosin as dye for rubber latex, polymethyl methacrylate sheets and paper. African Journal of Biotechnology 12 (17)

Kumar A, Vishwakarma HS, Singh J, Dwivedi S, Kumar M (2015) Microbial pigments: production and their applications in various industries. Int J Pharm Chem Biol Sci 5 (1):203-212

Lam KS (2006) Discovery of novel metabolites from marine actinomycetes. Current opinion in microbiology 9 (3):245-251

Lee D-R, Lee S-K, Choi B-K, Cheng J, Lee Y-S, Yang SH, Suh J-W (2014) Antioxidant activity and free radical scavenging activities of Streptomyces sp. strain MJM 10778. Asian Pacific journal of tropical medicine 7 (12):962-967

Lee JS, Kim Y-S, Park S, Kim J, Kang S-J, Lee M-H, Ryu S, Choi JM, Oh T-K, Yoon J-H (2011) Exceptional production of both prodigiosin and cycloprodigiosin as major metabolic constituents by a novel marine bacterium, Zooshikella rubidus S1-1. Applied and environmental microbiology 77 (14):4967-4973

Li D, Liu J, Wang X, Kong D, Du W, Li H, Hse C-Y, Shupe T, Zhou D, Zhao K (2018) Biological potential and mechanism of prodigiosin from Serratia marcescens subsp. lawsoniana in human choriocarcinoma and prostate cancer cell lines. International journal of molecular sciences 19 (11):3465

Long ITV (2004) Process for production of carotenoids, xanthophylls and apo-carotenoids utilizing eukaryotic microorganisms. Google Patents,

Mahon CR, Lehman DC (2022) Textbook of diagnostic microbiology-e-book. Elsevier Health Sciences,

Majumdar S, Priyadarshinee R, Kumar A, Mandal T, Mandal DD (2019) Exploring Planococcus sp. TRC1, a bacterial isolate, for carotenoid pigment production and detoxification of paper mill effluent in immobilized fluidized bed reactor. Journal of Cleaner Production 211:1389-1402

Mumtaz R, Bashir S, Numan M, Shinwari ZK, Ali M (2019) Pigments from soil bacteria and their therapeutic properties: A mini review. Current Microbiology 76:783-790

Namazkar S, Garg R, Ahmad WZ, Nordin N (2013) Production and characterization of crude and encapsulated prodigiosin pigment. International Journal of Chemical Sciences and Applications 4 (3):116-129

Narsing Rao MP, Xiao M, Li W-J (2017) Fungal and bacterial pigments: secondary metabolites with wide applications. Frontiers in microbiology 8:1113

Nawaz A, Chaudhary R, Shah Z, Dufossé L, Fouillaud M, Mukhtar H, ul Haq I (2020) An overview on industrial and medical applications of bio-pigments synthesized by marine bacteria. Microorganisms 9 (1):11

Özkan G, Bilek SE (2014) Microencapsulation of natural food colourants. International Journal of Nutrition and Food Sciences 3 (3):145-156

Palanichamy V, Hundet A, Mitra B, Reddy N (2011) Optimization of cultivation parameters for growth and pigment production by Streptomyces spp. isolated from marine sediment and rhizosphere soil. International Journal of plant, animal and environmental sciences 1 (3):158-170

Pandey B, Ghimire P, Agrawal V (2004) Studies on the antimicrobial activity of actinomycetes isolated from Khumbu region of Nepal. Retrieved on 21

Panesar R, Kaur S, Panesar PS (2015) Production of microbial pigments utilizing agro-industrial waste: a review. Current Opinion in Food Science 1:70-76

Parmar F, Kushawaha N, Highland H, George L-B (2015) In vitro antioxidant and anticancer activity of Mimosa pudica linn extract and L-Mimosine on lymphoma daudi cells. Cancer cell 1:100-104

Patil S, Paradeshi J, Chaudhari B (2016) Anti-melanoma and UV-B protective effect of microbial pigment produced by marine Pseudomonas aeruginosa GS-33. Natural Product Research 30 (24):2835-2839

Prabhu S, Arun P, Rekha A (2013) Evaluation of antioxidant activity of carotenoid isolated from Fontibacter flavus YUAB-SR-25. Int J Nat Prod Res 3:62-67

Prakash A, Rigelhof F, Miller E (2001) Medallion laboratories analytical progress: Antioxidant activity. J DeVries, PhD (ed), Medallion Laboratories 19 (2):1-6

Priya KA, Satheesh S, Ashokkumar B, Varalakshmi P, Selvakumar G, Sivakumar N (2013) Antifouling activity of prodigiosin from estuarine isolate of Serratia marcescens CMST 07. Microbiological research in agroecosystem management:11-21

Ramesh C, Vinithkumar NV, Kirubagaran R, Venil CK, Dufossé L (2020) Applications of prodigiosin extracted from marine red pigmented bacteria Zooshikella sp. and actinomycete Streptomyces sp. Microorganisms 8 (4):556

Ravikumar S, Uma G, Gokulakrishnan R (2016) Antibacterial property of halobacterial carotenoids against human bacterial pathogens.

Rezaeeyan Z, Safarpour A, Amoozegar MA, Babavalian H, Tebyanian H, Shakeri F (2017) High carotenoid production by a halotolerant bacterium, Kocuria sp. strain QWT-12 and anticancer activity of its carotenoid. EXCLI journal 16:840

Rinkal C, Srivathsa N (2018) Actinomycetes: a general review. International Journal for Research in Applied Science and Engineering Technology 6 (3):1267-1273

Romanenko L, Tanaka N, Svetashev V, Kalinovskaya N, Mikhailov V (2015) Rheinheimera japonica sp. nov., a novel bacterium with antimicrobial activity from seashore sediments of the Sea of Japan. Archives of microbiology 197:613-620

Rossi F, De Philippis R (2016) The physiology of microalgae. Developments in Applied Phycology 6:565-590

Saha S, Thavasi R, Jayalakshmi S (2008) Phenazine pigments from Pseudomonas aeruginosa and their application as antibacterial agent and food colourants. Res J Microbiol 3 (3):122-128

Sarkar G, Suthindhiran K (2022) Diversity and biotechnological potential of marine actinomycetes from India. Indian Journal of Microbiology 62 (4):475-493

Sen T, Barrow CJ, Deshmukh SK (2019) Microbial pigments in the food industry—challenges and the way forward. Frontiers in nutrition 6:7

Sharma R, Chandan G, Chahal A, Saini RV (2017) Antioxidant and anticancer activity of methanolic extract from Stephania elegans. Int J Pharm Pharm Sci 9 (2):245-249

Sibero MT, Bachtiarini TU, Trianto A, Lupita AH, Sari DP, Igarashi Y, Harunari E, Sharma AR, Radjasa OK, Sabdono A (2019) Characterization of a yellow pigmented coral-associated bacterium exhibiting anti-Bacterial Activity Against Multidrug Resistant (MDR) Organism. The Egyptian Journal of Aquatic Research 45 (1):81-87

Silva PI, Stringheta PC, Teófilo RF, De Oliveira IRsN (2013) Parameter optimization for spray-drying microencapsulation of jaboticaba

(Myrciaria jaboticaba) peel extracts using simultaneous analysis of responses. Journal of Food Engineering 117 (4):538-544

Smith AW, Camara-Artigas A, Olea C, Francisco WA, Allen JP (2004) Crystallization and initial X-ray analysis of phenoxazinone synthase from Streptomyces antibioticus. Acta Crystallographica Section D: Biological Crystallography 60 (8):1453-1455

Spada JC, Noreña CPZ, Marczak LDF, Tessaro IC (2012) Study on the stability of β-carotene microencapsulated with pinhão (Araucaria angustifolia seeds) starch. Carbohydrate polymers 89 (4):1166-1173

Srilekha V, Krishna G, Mahender P, Charya MS (2018) Investigation of in vitro cytotoxic activity of pigment extracted from Salinicoccus sp. isolated from Nellore sea coast. Journal of Marine Medical Society 20 (1):31-33

Srilekha V, Krishna G, Srinivas VS, Charya MS (2017) Antimicrobial evaluation of bioactive pigment from Salinicoccus sp. isolated from Nellore sea coast. Int J Biotechnol Biochem 13:211-217

Takano H, Asker D, Beppu T, Ueda K (2006) Genetic control for light-induced carotenoid production in non-phototrophic bacteria. Journal of Industrial Microbiology and Biotechnology 33 (2):88-93

Van Duin D, Paterson DL (2016) Multidrug-resistant bacteria in the community: trends and lessons learned. Infectious disease clinics 30 (2):377-390 van Wezel GP, McKenzie NL, Nodwell JR (2009) Applying the genetics of secondary metabolism in model actinomycetes to the discovery of new antibiotics. Methods in enzymology 458:117-141

Venegas FA, Köllisch G, Mark K, Diederich WE, Kaufmann A, Bauer S, Chavarría M, Araya JJ, García-Piñeres AJ (2019) The bacterial product violacein exerts an immunostimulatory effect via TLR8. Scientific Reports 9 (1):13661

Venil CK, Zakaria ZA, Ahmad WA (2013) Bacterial pigments and their applications. Process Biochemistry 48 (7):1065-1079

Wagh P, Mane R (2017) Identification and characterization of extracellular red pigment producing Neisseria spp. isolated from soil sample. Int J Innov Knowl Concept 5:23-25

Wang Y, Nakajima A, Hosokawa K, Soliev AB, Osaka I, Arakawa R, Enomoto K (2012) Cytotoxic prodigiosin family pigments from Pseudoalteromonas sp. 1020R isolated from the Pacific coast of Japan. Bioscience, biotechnology, and biochemistry 76 (6):1229-1232

Wang Z, Tschirhart T, Schultzhaus Z, Kelly E, Chen A, Oh E, Nag O, Glaser E, Kim E, Lloyd P (2019) Characterization and application of melanin produced by the fast-growing marine bacterium Vibrio natriegens through heterologous biosynthesis. Appl Environ Microbiol 86:02749-02719

Yi J, Lam TI, Yokoyama W, Cheng LW, Zhong F (2014) Cellular uptake of β-carotene from protein stabilized solid lipid nanoparticles prepared by homogenization–evaporation method. Journal of agricultural and food chemistry 62 (5):1096-1104

Young AJ, Lowe GM (2001) Antioxidant and prooxidant properties of carotenoids. Archives of Biochemistry and biophysics 385 (1):20-27

Zeng Z, Guo XP, Cai X, Wang P, Li B, Yang JL, Wang X (2017) Pyomelanin from Pseudoalteromonas lipolytica reduces biofouling. Microbial biotechnology 10 (6):1718-1731

# CHAPTER III

# Archaeal Pigments: Synthesis Mechanisms and Multifaceted Applications

**Ghada Hegazy,** *PhD*

National Institute of Oceanography and Fisheries, NIOF, Cairo, Egypt.

## Abstract

Carotenoids are natural pigments that have diverse functions and potential health benefits. They play a crucial role in photosynthesis and are found in various organisms, including bacteria, cyanobacteria, algae, and plants. Carotenoids and their derivatives, known as apocarotenoids, exhibit a wide range of biological activities, such as antioxidant, anti-inflammatory, anti-cancer, and neuroprotective properties. Due to their color and health-promoting properties, carotenoids are widely used in the food, pharmaceutical, nutraceutical, and cosmetic industries. Over 1200 naturally occurring carotenoids and apocarotenoids have been identified so far, with different structures and functional groups. Haloarchaea, a group of halophilic archaea, are known for their ability to produce carotenoids. However, research on carotenoid production by archaea, particularly Haloarchaea, is limited compared to other microorganisms. Bacterioberin is the most abundant carotenoid found in Haloarchaea, and other carotenoids such as phytoene, lycopene, and β-carotene are also present in lower concentrations. Raman spectroscopy is a valuable tool for identifying characteristic carotenoids in Haloarchaea.

**Keywords:** Carotenoids, natural pigments, antioxidant activity, Haloarchaea, bacterioruberin

## Introduction

Carotenoids are natural pigments that have received special attention due to ecophysiological functions, biotechnological applications, and their beneficial potential effects on human health (Gagez et al., 2012; Pajot et al., 2022). The color of these molecules can vary from colorless to red, through different tones of yellow-orange and they represent the second rich natural pigment in nature (Yatsunami, et al., 2014; Nisar et al., 2015). For example, fucoxanthin, an epoxy carotenoid found in microalgae and brown algae, is the main carotenoid present in marine ecosystems, accounting for 10% of the total carotene produced (Pajot et al., 2022). Apocarotenoids, for example retinal, defined as derivatives of carotenoids that are produced by chemical or enzymatic methods of oxidative cleavage.

The most important physiological function of carotenoids is photosynthesis, because they are actively involved in the harvesting of photons in the light-harvesting complexes of photosynthetic organisms. Therefore, they are found in all phototrophic bacteria, cyanobacteria, algae, and terrestrial plants. Their biological activities have been studied in detail in photosynthetic organisms and include other important ecological physiological functions such as membrane stabilizing activity, photoprotection, and anti-oxidation. animals, including humans, consume phototrophs and use carotenoids or apocarotenoids for important physiological functions including body color, sexual attractiveness, vitamin production, antioxidant activity, activation of transcription factors, and photoreceptors. Pharmacological studies have also demonstrated different biological activities of carotenoids and apocarotenoids, including antimicrobials, antifungals, antidiabetics, immunostimulants, anti-obesity, anti-inflammatory, anti-cancer and prevent cancer, anti-metastasis, anti-angiogenic, radioprotective, anti-atherosclerotic, neuroprotective and chemosensitive properties of multidrug-resistant cancer cells (De Oliveira et al., 2018; Pasquet et al., 2011; Ávila-Román et al., 2021; Lau & Kwan, 2022).

Because of their color and health-promoting properties, carotenoids are found in many varieties of applications in the food, pharmaceutical, nutraceutical, and cosmetic industries (Sandmann, 2014). To date in 2022, 1204 naturally occurring carotenoids and apocarotenoids have been described in three areas of life. These include 324 carotenes found in bacteria (from including 262 types found only in bacteria), 680 types of carotenoids in eukaryotes (including 621 eukaryotes only), and 25 Archaea carotenoids (including 13 found only in archaea). found in archaea) (Figure 1).

Carotenoid and apocarotenoid bacteria are synthesized by photosynthetic bacteria, cyanobacteria, and non-photosynthetic bacteria. All or most Archaea synthesize carotenoids. Eukaryotes carotenoids and apocarotenoids are present in photosynthetic carotenoids (microalgae, algae, terrestrial and marine plants), non-photosynthetic carotenes (fungi) and non-carotene non-photosynthetic species (animals and humans) obtain them through the consumption of productive species. Given this biodiversity, the number of species containing and/or synthesizing carotenoids and apocarotenoids is very large and cannot be precisely identified.

The carotene family consists mainly of the hydrocarbon backbone molecules C40 (1121 molecules) but also includes a smaller number of molecules containing 30 (37 carotenoids), 35 (5 carotenes), 45 (13 carotenes), and 50 carbon atoms (33 molecules). C40 Carotenoids are formed by the polymerization of 8 isoprene units, often modified by different oxidizing groups to obtain cyclic or open-chain xanthophylls.

Thus, their polarity can vary from very hydrophobic to amphoteric or relatively polar. All carotenoids possess a long chain of conjugated double bonds and a near-bilateral symmetry around the central double bond as common chemical characteristics (Rao & Rao, 2007). According to without or in the presence of an oxygen atom, carotenoids are essentially classified as carotenes or carotene hydrocarbons, consisting only of carbon and hydrogen (for example, lycopene and) β,β-carotene) and oxidizing xanthophylls or carotenoids may contain epoxy, carbonyl, hydroxyl, methoxy or carboxylic acid functional groups (e.g. lutein, canthaxanthin, zeaxanthin, violaxanthin, capsorubin, fucoxanthin and astaxanthin) (Rivera & Canela-Garayoa, 2012; Maoka, 2020). Some xanthophylls exist as esters of fatty acids, glycosides, and sulfates and can be found bound to proteins. complexes like rhodopsin. Figure 2 presents the 2D structures of two model carotenoids, carotene, and xanthophyll.

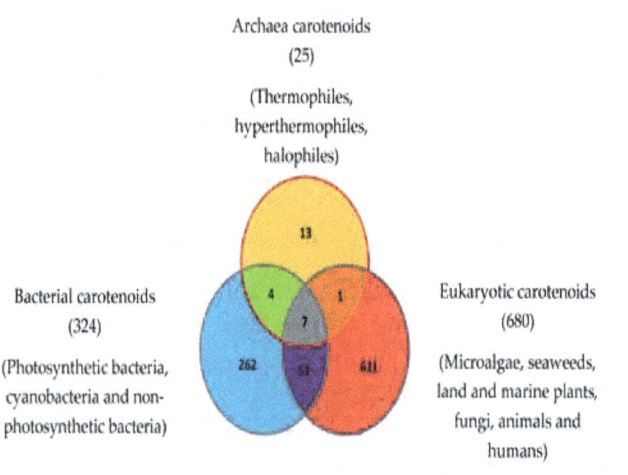

**Figure 2** Carotenoids and apocarotenoids in the three domains of life (Coulombier et al., 2021).

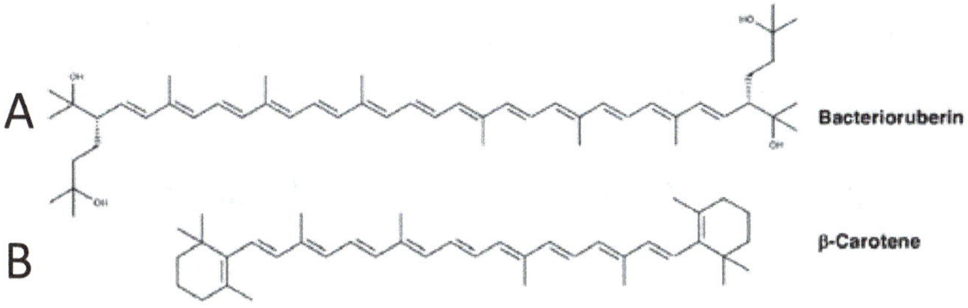

**A** — Bacterioruberin

**B** — β-Carotene

**Figure 3** 2D structures of model carotenoids, (A) β, β -carotene and (B) zeaxanthin (Coulombier et al., 2021).

- **Types, contents, and biosynthesis of halophilic carotenoids**

Halophilic archaea are predominantly halophilic microorganisms classified in the family Haloferacaceae, Phylum Euryarchaeota, domain Archaea. They are (mostly) aerobic generally colored red due to the presence of carotenoids. They form the predominant microbial community in highly halophilic environments. To survive in these extreme habitats, they have developed several strategies.(i) amino acid residues are predominant on the surface of halophilic proteins; (ii) cells accumulate high intracellular KCl concentrations to cope with high ionic strength or osmolytes such as 2-sulfotrehalose (Desmarais et al., 1997) ;(ii) cell bilayers have different compositions, structures, etc.; These adaptations have made Haloarchaea an excellent and innovative source for a variety of molecules of great interest in biotechnology such as enzymes that can be activated at high temperature and high ionic strength (Madern et al., 2004; Bonete et al., 2011) PHB, PHA and carotenoids.

Regarding carotenoids, information in the literature on the carotenoid profiles of extremophiles is scarce compared to that available from other microorganisms, and little literature focuses on carotenoid production by archaea in general, and Haloarchaea in particular. It is important to emphasize that despite the large number of publications on this topic, only 1.3% of them are related to haloarchaea carotenoids (780 articles on haloarchaea carotenoids compared to 61590 articles on carotenoids in general). The literature clearly supports that most members of the Haloferaceae family contain bacterioberin (the most abundant C50 in most analyzed haloarchaea species) and its precursor (2-isopentenyl-3,4-dehydro It is capable of synthesizing C50 carotenoids, including rhodopin (IDR)), bisanhydrobacterioverin (BABR) and monoanhydrobacterioverin (MABR)) (Kelly et al., 1967; Kushwaha et al., 1975). Several others, including 3,4-dehydromonoanhydrobacterioverin, haloxanthin (a derivative of the previous one with peroxide end groups), and 3,4-epoxymonoanhydrobacterioverin, identified in *Haloferax volcanii*. was found in small amounts (Ronnekleiv et al., 1995; Bidle et al., 2007). Other carotenoids such as phytoene, lycopene, and β-carotene are also produced by these species, but at lower concentrations (Goodwin et al., 1980). These carotenoids are found in cell membranes and are responsible for the color of red colonies when Haloarchaea cells are grown on solid media, or the red (mainly summer) color exhibited by salty coastal ponds. Indeed, the level of bacterioberin pigments in biomass has been used to monitor the density of halophilic archaeal communities in halophilic environments (Oren & Gurevich., 1995).

Other carotenoids have been identified at very low concentrations in halophilic archaea. Lycopersene, cis and trans phytoene, cis and trans phytofluene, neo-β-carotene and neo-α-carotene. The low concentrations of these compounds suggest that they can be used as precursors for the synthesis of other carotenoids, including lycopene, retinal, and members of the bacterioverin group. Some species can also produce the ketocarotenoid canthaxanthin in addition to other carotenoids (Yatsunami, 2014). Although this is a common carotenoid profile exhibited by most Haloarchaea species, it is important to note that some of them can produce large amounts of cantanxanthin, β-carotene, and trans-astaxanthin (Asker et al., 2002).

The presence of characteristic carotenoids (α-bacteriobins and derivatives) in Haloarchaea cells is readily identified by Raman spectroscopy (Oren, 2004; Jehlička et al., 2014). Thanks to this technique, α-bacterioruberin was identified as the major carotenoid of the haloarchaea. *Halobacterium salinarum* strains

NRC-1 and R1, *Halorubrum sodomense, Haloarcula vallismortis, Haloarcula japonica* (68.1% (mol%) of total carotenoids (Yatsunami, 2014). bisanehydrobactriolberin (9.3%) and isopentenyldehydrorhodopine (<0.1%) [5]. The total carotenoid content of *Haloarcula japonica* was 335 µg g-1 dry matter, while that of *Halobacterium salinarum* and *Halococcus morrhuae* was 89 and 45 µg g-1, respectively (Mandelli et al., 2012). Carotenoid biosynthesis in haloarchaea was first studied in the late 1970s. At that time, it was found that the biosynthetic pathway of his C40 carotene in *Halobacterium* was as follows:

Isopentenyl pyrophosphate produces trans-phytoene, trans-phytofluene, ζ-carotene, neurospores, lycopene, gamma-carotene, and finally β - carotenes. This pathway differs from that of higher plants in that the cis isomers of phytoene and phytofluene are not in the main pathway of carotene biosynthesis as in higher plants, bacterioruberin, on the other hand, is proposed to be synthesized by adding C5 isoprene units to both ends of the lycopene chain, followed by the introduction of four hydroxyl groups. Evidence supporting these proposals was reported about 40 years ago from experiments using nicotine to inhibit bacterioruberin synthesis (Kushwaha & Kates., 1976; Kushwaha & Kates., 1979). The presence of multiple genes at multiple stages in carotenoid production in *Halobacterium* NRC-1 suggests that multiple biosynthetic pathways may exist (Peck et al., 2001; Dassarma et al., 2001). Computational genome and pathway analyzes of halophilic archaea performed by Falb and coworkers suggested that phytoene is reduced to lycopene by phytoene desaturases. Lycopene is the junction point for the synthesis of bacterioruberin (C50) and β-carotene (C40). Although the reaction from lycopene to bacterioruberin is not yet fully understood, there is some evidence that lycopene cyclase (OE3983R) converts lycopene to β-carotene in *Halobacterium salinarum* str. NRC-1. A recent study in *Haloarcula japonica* identified genes named c0507, c0506, and c0505 for carotenoid 3,4-desaturase (CrtD), bifunctional lycopene elongase and 1,2-hydratase (LyeJ), and C50 carotenoid 2. has been positively confirmed to produce "-Hydratase (CruF). The three carotenoid biosynthetic enzymes mentioned above catalyze the reaction that converts lycopene to bacterioruberin in *Haloarcula japonica* (Yang et al., 2015). Figure 3 compares carotenoid biosynthesis in photosynthetic organisms and Haloarchaea (based on results reported from *Halobacterium*, *Haloarcula*, and preliminary evidence from *Haloferax* genome analysis). This pigment protects archaeal cells from damage caused by high light intensities in the visible and ultraviolet regions of the spectrum and aids photorecovery. It is also involved in strengthening cell membranes. It was first described from cells of *Halobacterium* species. The biosynthesis of a common C50 carotenoid and the effects of several compounds on this biosynthesis were first described from *Halobacterium cutirubrum* (Halobacteriaceae family). A few years later, it was reported that bacterioruberin was synthesized from other C50 carotenoids such as isopentenyldehydrorhodopine, bisanehydrobacterioberin and monoanhydrobacterioberin (Yatsunami, 2014). (ii) osmotic pressure; (iii) presence of various compounds such as aniline (Figure 3 and 4).

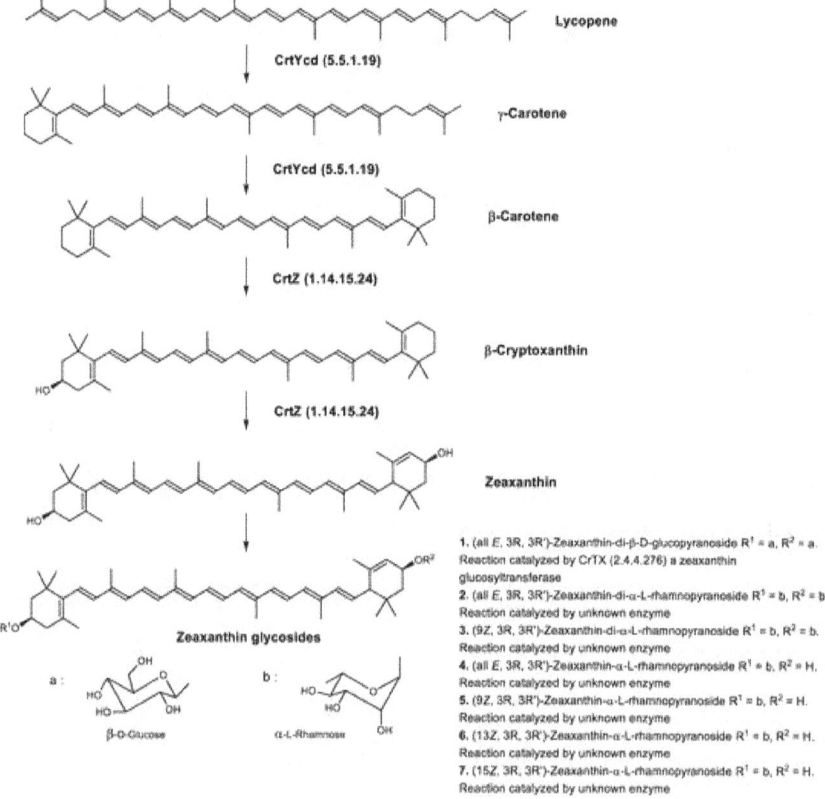

**Figure 3** Carotenoids biosynthesis pathway in *Haloarcula* (Yang et al., 2015).

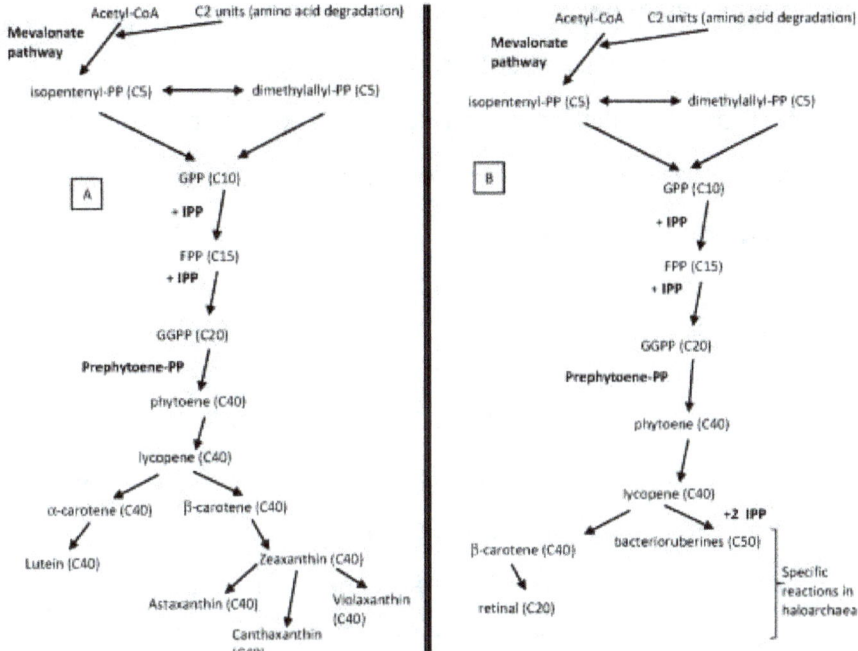

**Figure 4** Comparison of isoprenoid biosynthesis in photosynthetic organisms (A) and the proposed biosynthetic pathway in haloarchaea (B) (Yang et al., 2015).

- **Potential of carotenoid production from halophilic archaea**

The growth culture conditions that are essential to stimulate fast growth of the halophilic archaea to produce large amounts of carotenoid include high NaCl concentrations ranging (from 20% to 30% w/v). However, the production of large amounts of carotenoids from halophilic archaea generally needs much lower NaCl concentrations or normally below 16% w/v. Such lower NaCl concentrations lead to slower rates of growth or even lysis of cells. Therefore, carotenoid production and growth of haloarchaea are often contradictory events. Moreover, besides the salinity of the culture medium, other factors as pH and temperature also may affect the production of carotenoids and the growth rates of halophilic archaea. In some cases, the conditions and the nutrient composition changes of the culture medium might lead to enhancement C50 carotenoid production (Tian et al., 2007; Fang et al., 2010).

Volumetric productivity is estimated in terms of $g \cdot L^{-1}/day^{-1}$, and the productivity of carotenoid can also be estimated per volume of the reactor volume ($g \cdot L^{-1}/day^{-1}$) or as $mg \cdot g^{-1}$ dry weight·/day$^{-1}$. At large scale production, the surface is involved as a factor carotenoids productivity. For example, the real productivity biomass is expressed by $g \cdot m^{-2}/day^{-1}$, the rate of produced biomass in time clearly determines the potential of carotenoid production by the halophilic archaea. no published data on both carotenoid and biomass productivities from halophilic archaea at large scales but there are few results on the laboratory scale. these data are needed to know the carotenoids production potential of the halophilic archaea. For example, the average time was needed for the cultivation of *Halorubrum* sp. and the other haloarchaea species to reach the exponential phase end in batch systems is about 11 days, the biomass dry weight resulted is about 0.8 $g \cdot L^{-1}$. These data of productivity should be higher when the haloarchaea were cultivated in continuous systems of production at the optimum growing conditions. Therefore, the haloarchaea should be taken a chance to be studied for assessing the carotenoid production potential (Figure 5) (Fang et al., 2010).

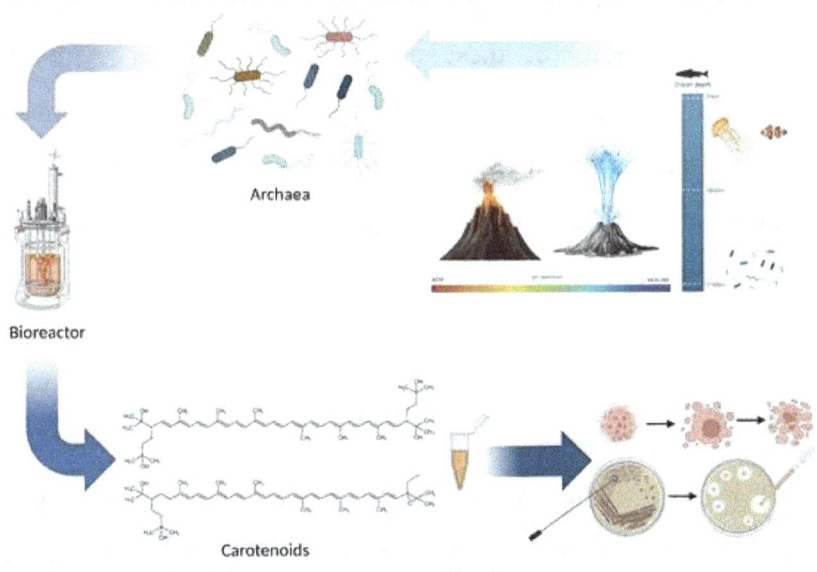

**Figure 5** Production of the carotenoids from archaea (Fang et al., 2010).

### Carotenoids Biological activities

Although the biological activities of the carotenoids which produced by haloarchaea, the potential benefits of these carotenoids on human and animal cells have not been examined yet. However, some evidence supporting the antioxidant activities of the haloarchaea carotenoids. For example, the halophilic archaea *Halobacterium salinarum* produce different types of carotenoids such as phytoene, lycopene and salinixanthin. These types of carotenoids have been examined for their free radical scavenging and antioxidant activity by DPPH (di(phenyl)-(2,4,6-trinitrophenyl) iminoazanium) assay and the results confirmed the antioxidant activity of these carotenoids. A further cytotoxic properties analysis against liver cancer cell lines of humans resulted increasing in dose-dependent of the carotenoids of cytotoxicity on these cell lines, suggesting the anti-cancer activity (Sikkandar et al., 2013).

- **Bacterioruberin as antioxidant**

The oxygen-reactive species (ROS) scavenging capacity is dependent on carotenoid concentration. On the other hand, the antioxidant capacity of carotenoids generally depends on carbon chain length, number of conjugated double bond pairs, and carotenoid concentration. Bacterioruberin contains 13 pairs of conjugated double bonds compared to 9 pairs of conjugated double bonds in β-carotene. Bacterioruberin is therefore far superior to β-carotene as a radical scavenger. Proven to protect cells from oxidative damage. As a result of this important biological role, Haloarchaea avoid lethal damage from intense light and resist oxidative DNA damage from radiography, UV irradiation, high-dose (5 kGy) gamma irradiation, and $H_2O_2$ exposure. It has been clearly established that carotenoids from halophilic microorganisms have higher antioxidant capacity than carotenoids produced by other microorganisms (either extremophilic or non-extremophilic) (Tian et al., 2007).

- **Membrane rigidity is controlled by bacterioruberin**

The four hydroxyl substituents of this C50 bipolar carotenoid suggest that bacterioruberin functions as a 'rivet' for membrane cells. This carotenoid has some effect on membrane fluidity, acts as a water barrier, allows permeation of oxygen and other molecules, and allows strains to survive in high-salinity or low-temperature conditions (Fang et al., 2010).

- **Bacterioruberin part of the rhodopsin complex**

Archaerhodopsin-2 (aR2) is a retinal protein-carotenoid complex found in the *Halorubrum* sp. membrane. It functions as a light-driven proton pump and is crucial for Haloarchaea cells to obtain energy. Using crystallographic studies, bacterioruberin has been shown to bind to clefts between subunits of the trimeric archaerhodopsin structure. Bacterioverin is also part of a complex composed of this carotenoid and halorhodopsin in the membranes of haloarchaea, such as those of *Natronomonas pharaonis*. Halorhodopsin is a retinal protein with seven transmembrane helices that functions as a light-driven inward Cl (-) pump (Table 1) (Sasaki et al., 2012).

**Table 1** Recent haloarchaeal carotenoids applications in biomedicine.

| Carotenoid source | Biological applications |
|---|---|
| *Halobacterium halobium*, *Halogeometricum limi* and *Haloplanus vescus* carotenoid extracts | Human hepatoma HepG2 cells viability decreasing. |
| Halophilic archaea carotenoids | Combination therapy with radiation treatment of reduction of solid tumor. oxidative stress protection. $H_2O_2$ protection in erythrocytes. Anticancer activity |
| *Haloferax volcanii*, *Halopelagius inordinatus*, *Halogranum rubrum*, and *Halogeometricum rufum* carotenoids extracts | Scavenging radicals' activity |
| *Haloferax volcanii* carotenoid extract | The viability of sperm cells beneficial effects |

**Biotechnological applications of halophilic archaea carotenoids**

Carotenoids have many uses as coloring agents (food and cosmetics), feed additives poultry, fish, livestock, crustaceans, antioxidant, antitumor, vitamin A precursor, heart disease prevention, in vitro antibody enhancer, manufacturing. Therefore, it is widely used in the food, pharmaceutical and cosmetic industries. As a coloring agent and functional ingredient (Torregrosa-Grespo et al., 2017).

New research shows potential for halophilic archaea as carotenoids production natural sources, thanks to the easy growth for carotenoid production by different culture conditions, genetic manipulation, and the possibility of downstream processes of the haloarchaeal cells to extract the carotenoids. The haloarchaea has unique characteristics allowing them to be them suitable carotenoids production potential sources, including: (i) the haloarchaea high-salt tolerance allows their cultivation under non-sterile conditions due to the ability of high salt concentrations to prevent the growth of other microorganisms. This unique feature allows advantageous cultivation of haloarchaea when compared to the other microorganisms cultivation; (ii) the process to extract the carotenoids is very simple because the cell lysis is induced in lower salt concentrations and in this case, the extraction of carotenoids could be easily conducted directly from the haloarchaea cells without needing any mechanical techniques which is used in case of other sources such as plants and (iii) the pigments extraction and purification procedures appear to be simpler than the procedures from other biological sources. Therefore, the production capacity of carotenoids from haloarchaea should be examined to evaluate alternative commercial sources for production of carotenoids (Torregrosa-Grespo et al., 2017).

## Conclusion

We have highlighted Archaea's potential in synthesizing Carotenoids can find various biotechnological and pharmaceutical applications. Archaea are interesting microorganisms for large-scale carotene production. medium scale, as cheap and fast-growing systems can be incorporated downstream easily through extraction and purification processes. Harsh growing conditions limit the risk of contamination by other microorganisms. The technique of haloarchaea is now possible thanks to the knowledge of the metabolic pathways of carotene generation, the sequence of the haloarchaea genome, and haloarchaea genetic manipulation. carotenoid production by haloarchaea can also be enhanced by optimizing the culture process environment in terms of salinity, pH, and temperature. Carotenoids produced by halophiles Archaea can play the dual role of membrane stabilizer and high antioxidant compounds, making them essential compounds for the survival of these microorganisms but also represent molecules with high potential for health, cosmetics, and biotechnology application. A small amount of research is devoted to the biosynthesis of natural products of archaea suggests that this area of research remains relatively unexplored and that more discoveries will be made in the coming years.

# References

Asker D., Awad T., Ohta Y. Lipids of *Haloferax. alexandrinus* strain TM$^T$: An extremely halophilic canthaxanthin-producing archaeon. *J. Biosci. Bioeng.* 2002;**93**:37–43. doi: 10.1016/S1389-1723(02)80051-2. [PubMed] [CrossRef] [Google Scholar].

Ávila-Román, J.; García-Gil, S.; Rodríguez-Luna, A.; Motilva, V.; Talero, E. Anti-Inflammatory and Anticancer Effects of Microalgal Carotenoids. Mar. Drugs 2021, 19, 531. [CrossRef].

Bidle K.A., Hanson T.E., Howell K., Nannen J. HMG-CoA reductase is regulated by salinity at the level of transcription in *Haloferax volcanii. Extremophiles.* 2007;**11**:49–55. doi: 10.1007/s00792-006-0008-3. [PubMed] [CrossRef] [Google Scholar].

Bonete M.J., Martínez-Espinosa R.M. Enzymes from Halophilic Archaea: Open Questions. In: Ventosa A., Oren A., editors. *Halophiles and Hypersaline Environments: Current Research and Future Trends.* Springer-Verlag GmbH; Berlin, Germany: 2011. pp. 358–370. [Google Scholar].

Coulombier, N.; Jauffrais, T.; Lebouvier, N. Antioxidant Compounds from Microalgae: A Review. Mar. Drugs 2021, 19, 549. [CrossRef].

Dassarma S., Kennedy S.P., Berquist B., Victor N.W., Baliga N.S., Spudich J.L., Krebs M.P., Eisen J.A., Johnson C.H., Hood L. Genomic perspective on the photobiology of *Halobacterium* species NRC-1, a phototrophic, phototactic, and UV-tolerant haloarchaeon. *Photosynth. Res.* 2001;**70**:3–17. doi: 10.1023/A:1013879706863. [PubMed] [CrossRef] [Google Scholar].

De Oliveira, R.G., Jr.; Adrielly, A.F.C.; da Silva Almeida, J.R.G.; Grougnet, R.; Thiéry, V.; Picot, L. Sensitization of Tumor Cells to

Chemotherapy by Natural Products: A Systematic Review of Preclinical Data and Molecular Mechanisms. Fitoterapia 2018, 129, 383–400. [CrossRef] [PubMed].

Desmarais D., Jablonski P.E., Fedarko N.S., Roberts M.F. 2-Sulfotrehalose, a novel osmolyte in haloalkaliphilic archaea. *J. Bacteriol.* 1997;**179**:3146–3153. [PMC free article] [PubMed] [Google Scholar].

Fang C.J., Ku K.L., Lee M.H., Su N.W. Influence of nutritive factors on $C_{50}$ carotenoids production by *Haloferax mediterranei* ATCC 33500 with two-stage cultivation. *Bioresour. Technol.* 2010;**101**:6487–6493. doi: 10.1016/j.biortech.2010.03.044. [PubMed] [CrossRef] [Google Scholar].

Gagez, A.-L.; Thiery, V.; Pasquet, V.; Cadoret, J.-P.; Picot, L. Epoxycarotenoids and Cancer. Review. Bioact. Compd. 2012, 8, 109–141. [CrossRef].

Goodwin T.W., Britton G. Distribution and analysis of carotenoids. In: Goodwin T.W., editor. *Plant Pigments.* Academic Press; London, UK: 1980. pp. 61–132. [Google Scholar].

Jehlička J., Edwards H.G., Oren A. Raman spectroscopy of microbial pigments. *Appl. Environ. Microbiol.* 2014;**80**:3286–3295. doi: 10.1128/AEM.00699-14. [PMC free article] [PubMed] [CrossRef] [Google Scholar] Yatsunami R., Ando A., Yang Y., Takaichi S., Kohno M., Matsumura Y., Ikeda H., Fukui T., Nakasone K., Fujita N., et al. Identification of carotenoids from the extremely halophilic archaeon *Haloarcula japonica. Front. Microbiol.* 2014;**5**:100–105. doi: 10.3389/fmicb.2014.00100. [PMC free article] [PubMed] [CrossRef] [Google Scholar].

Kelly M., Jensen S.L. Bacterial carotenoids. XXVI. $C_{50}$-carotenoids. 2. Bacterioruberin. *Acta Chem.*

*Scand.* 1967;**21**:2578–2580. doi: 10.3891/acta.chem.scand.21-2578. [PubMed] [Crossref] [Google Scholar].

Kushwaha S.C., Kates M. Effect of glycerol on carotenogenesis in the extreme halophile, *Halobacterium cutirubrum. Can. J. Microbiol.* 1979;**25**:1288–1291. doi: 10.1139/m79-203. [PubMed] [Crossref] [Google Scholar].

Kushwaha S.C., Kates M. Effect of nicotine on biosynthesis of $C_{50}$ carotenoids in *Halobacterium cutirubrum. Can. J. Biochem.* 1976;**54**:824–829. doi: 10.1139/o76-118. [PubMed] [Crossref] [Google Scholar].

Kushwaha S.C., Kramer J.K., Kates M. Isolation and characterization of $C_{50}$-carotenoid pigments and other polar isoprenoids from *Halobacterium cutirubrum. Biochim. Biophys. Acta.* 1975;**398**:303–314. doi: 10.1007/s00792-006-0008-3. [PubMed] [Crossref] [Google Scholar].

Lau, T.-Y.; Kwan, H.-Y. Fucoxanthin Is a Potential Therapeutic Agent for the Treatment of Breast Cancer. Mar. Drugs 2022, 20, 370. [CrossRef].

Madern D., Camacho M., Rodríguez-Arnedo A., Bonete M.J., Zaccai G. Salt-dependent studies of NADP-dependent isocitrate dehydrogenase from the halophilic archaeon *Haloferax volcanii. Extremophiles.* 2004;**8**:377–384. doi: 10.1007/s00792-004-0398-z. [PubMed] [Crossref] [Google Scholar].

Mandelli F., Miranda V.S., Rodrigues E., Mercadante A.Z. Identification of carotenoids with high antioxidant capacity produced by extremophile microorganisms. *World J. Microbiol. Biotechnol.* 2012;**28**:1781–1790. doi: 10.1007/s11274-011-0993-y. [PubMed] [Crossref] [Google Scholar].

Maoka, T. Carotenoids as Natural Functional Pigments. J. Nat. Med. 2020, 74, 1–16. [CrossRef] [PubMed].

Nisar, N.; Li, L.; Lu, S.; Khin, N.C.; Pogson, B.J. Carotenoid Metabolism in Plants. Mol. Plant 2015, 8, 68–82. [CrossRef] [PubMed].

Oren A. Halophilic archaea on Earth and in space: Growth and survival under extreme conditions. *Philos. Trans. A Math. Phys. Eng. Sci.* 2014;**13**:372. doi: 10.1098/rsta.2014.0194. [PubMed] [CrossRef] [Google Scholar].

Oren A., Gurevich P. Dynamics of a bloom of halophilic archaea in the Dead Sea. *Hydrobiologia*. 1995;**315**:149–158. doi: 10.1007/BF00033627. [CrossRef] [Google Scholar]

Pajot, A.; Hao Huynh, G.; Picot, L.; Marchal, L.; Nicolau, E. Fucoxanthin from Algae to Human, an Extraordinary Bioresource: Insights and Advances in up and Downstream Processes. Mar. Drugs 2022, 20, 222. [CrossRef].

Pasquet, V.; Morisset, P.; Ihammouine, S.; Chepied, A.; Aumailley, L.; Berard, J.-B.; Serive, B.; Kaas, R.; Lanneluc, I.; Thiery, V.; et al. Antiproliferative Activity of Violaxanthin Isolated from Bioguided Fractionation of Dunaliella Tertiolecta Extracts. Mar. Drugs 2011, 9, 819–831. [CrossRef].

Peck R.F., Echavarri-Erasun C., Johnson E.A., Ng W.V., Kennedy S.P., Hood L., DasSarma S., Krebs M.P. *brp* and *blh* are required for synthesis of the retinal cofactor of bacteriorhodopsin in *Halobacterium salinarum. J. Biol. Chem.* 2001;**276**:5739–5744. doi: 10.1074/jbc.M009492200. [PubMed] [CrossRef] [Google Scholar].

Rao, A.; Rao, L. Carotenoids and Human Health. Pharmacol. Res. 2007, 55, 207–216. [CrossRef] [PubMed].

Rivera, S.M.; Canela-Garayoa, R. Analytical Tools for the Analysis of Carotenoids in Diverse Materials. J. Chromatogr. A 2012, 1224, 1–10. [CrossRef] [PubMed].

Ronnekleiv M., Liaaen-Jensen S. Bacterial Carotenoids 53*, $C_{50}$-Carotenoids 23; Carotenoids of *Haloferax volcanii versus* other Halophilic Bacteria. *Biochem. Syst. Ecol.* 1995;**23**:627–734. doi: 10.1016/0305-1978(95)00047-X. [Crossref] [Google Scholar].

Sandmann, G. Carotenoids of Biotechnological Importance. In Biotechnology of Isoprenoids; Schrader, J., Bohlmann, J., Eds.; Advances in Biochemical Engineering/Biotechnology; Springer International Publishing: Cham, Germany, 2014; Volume 148, pp. 449–467. ISBN 978-3-319-20106-1.

Sasaki T., Razak N.W., Kato N., Mukai Y. Characteristics of halorhodopsin-bacterioruberin complex from *Natronomonas pharaonis* membrane in the solubilized system. *Biochemistry.* 2012;**51**:2785–2794. doi: 10.1021/bi201876p. [PubMed] [Crossref] [Google Scholar].

Sikkandar, S.; Murugan, K.; Al-Sohaibani, S.; Rayappan, F.; Nair, A.; Tilton, F. Halophilic bacteria-A potent source of carotenoids with antioxidant and anticancer potentials. J. Pure Appl. Microbiol. 2013, 7, 2825–2830.

Tian B., Xu Z., Sun Z., Lin J., Hua Y. Evaluation of the antioxidant effects of carotenoids from *Deinococcus radiodurans* through targeted mutagenesis, chemiluminescence, and DNA damage analyses. *Biochim. Biophys. Acta.* 2007;**1770**:902–911. doi: 10.1016/j.bbagen.2007.01.016. [PubMed] [Crossref] [Google Scholar]

Torregrosa-Grespo, J., Galiana, P. C. & Espinosa, R. M. Biocompounds from Haloarchaea and their uses in biotechnology. J. Mar. Drugs. 4, 1–21 (2017).

Yang Y., Yatsunami R., Ando A., Miyoko N., Fukui T., Takaichi S., Nakamura S. Complete Biosynthetic Pathway of the $C_{50}$ Carotenoid Bacterioruberin from Lycopene in the extremely halophilic archaeon *Haloarcula* *japonica*. *J. Bacteriol.* 2015;**197**:1614–1623. doi: 10.1128/JB.02523-14. [PMC free article] [PubMed] [CrossRef] [Google Scholar].

Yatsunami, R.; Ando, A.; Yang, Y.; Takaichi, S.; Kohno, M.; Matsumura, Y.; Ikeda, H.; Fukui, T.; Nakasone, K.; Fujita, N.; et al. Identification of Carotenoids from the Extremely Halophilic Archaeon Haloarcula japonica. Front. Microbiol. 2014, 5, 100. [CrossRef].

# CHAPTER IV
# Fungal Pigments: Biosynthesis Pathways and Versatile Applications

**Khouloud M. Barakat,** *PhD*

National Institute of Oceanography and Fisheries, NIOF, Cairo, Egypt.

## Abstract

All taxonomic groups of fungi include carotenoids, also known as colored terpenoids, which vary depending on the chemical structures within each strain of the fungus. The primary carotenoids generated by fungus, including β-carotene, astaxanthin, torulene, and torularhodin, possess antioxidant properties that could prolong their survival in their native habitat. Additionally, a great deal of research has been done on the biosynthesis pathway of fungal carotenoids through the metabolic reaction of apocarotenoids by oxidative cleavage. Therefore, the current understanding of fungal-specific carotenoids, their distribution across fungal taxonomic groupings, and their production and conversion into simple acids is the main focus of this chapter. Because of their biology and the genetics underlying their synthesis process, fungal carotenoids have drawn particular interest. In conclusion, this chapter discusses the industrial significance of these carotenoids in the most developed yeasts and fungi. In conclusion, this chapter provides an overview of the industrial significance of these carotenoids in the most developed strains of yeast and fungus. It also highlights strategies and efforts in genetic engineering to improve or create pathways for the production of different fungal carotenoids for industrial or therapeutic purposes, ultimately leading to biotechnological potential as production systems.

**Keywords:** Fungal Carotenoids; Carotenogenic Pathways, Carotenoid Biosynthesis, Carotenoid Pathway Engineering.

## Introduction

Fungal carotenoids are a natural, safe pigment with many benefits, including beautiful, eye-catching colours, stability against light, heat, and pH, good quality, low cost, environmental friendliness, and weather independence. They are also easily extracted from growth media and have a wide range of uses (Ahmad et al., 2014; Wang et al., 2017; Zhao et al., 2019).

Authors will concentrate on the physical and chemical properties of fungal carotenoids, their classifications and derivatives, the biosynthetic pathway, their function and importance inside fungal cells, their bioactivity for the prevention and treatment of many diseases, the benefits of carotenoids production by fungal and yeast fermentation for industrial applications, and genetic strategies to develop carotenoids productivity on industrial scales for dealing with the challenges that the world is currently facing.

## Fungal carotenoids classification

About 700 to 800 derivatives of natural carotenoids are categorised into two main categories: non-oxygenated polyunsaturated hydrocarbon derivatives of carotenoids and oxygenated carotenoids (xanthophylls), all of which have the carbon number $C_{40}$. Thirty-four derivatives of fungal carotenoids, including twenty-one xanthophyll derivatives and thirteen non-oxygenated derivatives, have been described (Malik et al., 2012; Venil et al., 2013; Kirti et al., 2014; Tuli et al., 2015; Kuczynska and Jemiola-Rzeminska, 2017; Erasun and Johnson, 2018) (Figure 1).

## Major carotenoids producing fungi

*Phaffia rhodozyma* (and its teleomorph *Xanthophyllomyces dendrorhous*) and *Rhodosporidium*, *Rhodotorula*, *Sporobolomyces*, and *Sporidiobolus* species are among the yeast that can synthesise carotenoids (Black et al., 2008). *Blakeslea trispora* species is the most significant mould (Metlicar et al., 2019), and endophytic fungi like *Penicillium citrinum* that produce beta-carotene have recently been researched (Nagal and Panda, 2022). Several marine yeast species that have been isolated produce carotenoids. Astaxanthin has been obtained, in particular, from *Xanthophyllomyces*, *Rhodotorula*, and *Phaffia* (Ambati et al., 2014). Although yeasts produce less astaxanthin than algae, they develop more quickly and require less cultivation process (Mata-Gómez et al., 2014).

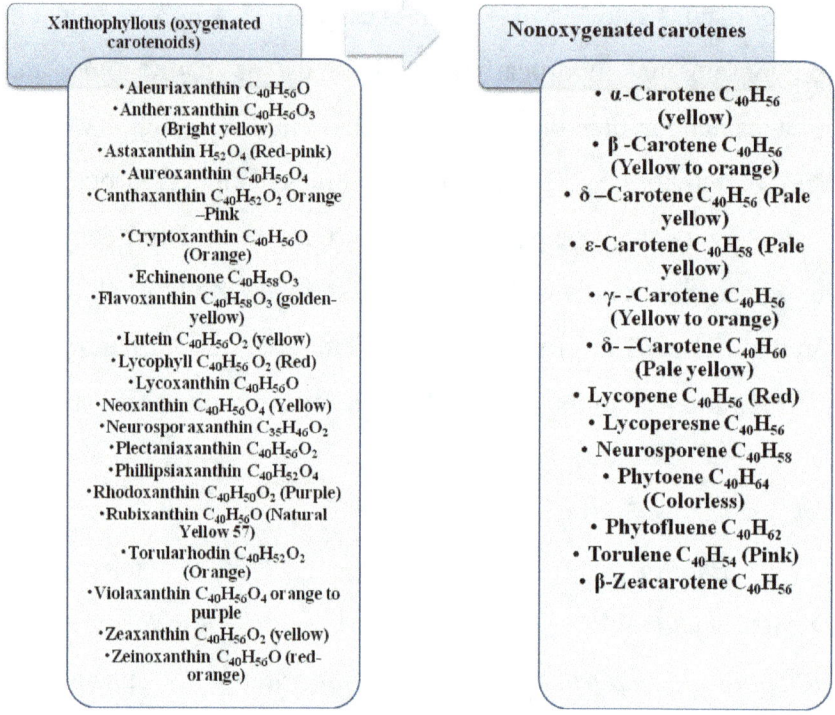

**Figure 1** Classification of 34 derivatives of carotenoids produced by fungi into two major groups based on their chemical structures and colors.

Carotenoids are produced even by protists that resemble fungi, including thraustochytrids. They are found in a variety of habitats, including the sediments of mangroves, estuaries, and deep-sea ecosystems, and have a wide geographic range, ranging from the polar to tropical regions (Raghukumar, 2002). In particular, *Thraustochytrium* strains ONC-T18 and CHN-1,

*Thraustochytriidae* sp. AS4-A1 (*Ulkenia* sp.) and *Aurantiochytrium* sp. KH105 synthesise various carotenoids including β-carotene, astaxanthin, zeaxanthin, cantaxanthin, phoenicoxanthin and echinenone (Aasen et al., 2016). As in the case of *Aurantiochytrium* sp. SK4 (Suen et al., 2014), engineering approaches have enabled an increase in the production of carotenoids (nine-fold enhanced astaxanthin content production). To increase our understanding of the sources of carotenoids for application in the biotechnological area, it is essential to develop genetic methods and genome sequencing applied to thraustochytrids.

- **β-Carotene**

The isoprenoid molecule -carotene has the molecular weight of 536.88 g/mol and the chemical formula $C_{40}H_{56}$. This compound's molecule is made up of two -ionone rings joined together by a polyene chain with nine conjugated double bonds. The compound has a maximum absorbance at 450 nm and is distinguished by a colour ranging from yellow to orange because of the structure of -carotene and the system of double bonds (Gul et al., 2015). Specific intestinal enzymes can transform one molecule of this chemical into two molecules of vitamin A, making -carotene the principal dietary supply of this vitamin (Grune et al., 2010). Yeast *Rhodotorula glutinis* (Bhosale and Gadre, 2002) and

*Rothia mucilaginosa* (Sharma and Ghoshal, 2020), and *Sporidiobolus pararoseus* (Manowattana et al., 2020) are effective producers of microbial -carotene.

- **Astaxanthin**

A member of the xanthophyll family is astaxanthin ($C_{40}H_{52}O_4$, 596.85 g/mol). A non-polar chain connects two polar -ionone rings. One hydroxyl group and one ketone group are present in each ring. The astaxanthin molecule contains a total of 13 double bonds, which is what gives this substance its potent antioxidant effects. Astaxanthin has ketone and hydroxyl groups, which enable esterification and define its polar nature. Astaxanthin is an optically active substance because it contains hydroxyl groups in the β-ionone rings. Astaxanthin has three isomers: enantiomers (3S, 3'S and 3R, 3'R) and a meso form (3R, 3'S) as a result of the presence of chiral centres at positions C-3 and C-3' (Jannel et al., 2020). *Xanthophyllomyces dendrorhous* is the principal fungus responsible for producing this substance, synthesizing (3R, 30R) isomer (Rodrguez-Sáiz et al., 2010).

- **Torulene**

The carotenes are made up of compounds like torulene ($C_{40}H_{54}$, 534.9 g/mol). One β-ionone ring and a polyene chain with twelve conjugated double bonds make up the torulene molecule. Its colour ranges from orange to orange-red depending on the concentration. *Rhodotorula* sp., *Sporidiobolus pararoseus*, and

*Neurospora* are the primary microorganisms that create torulene (Di Mascio et al., 1991; Wei et al., 2019). Torulene exhibits antioxidant and anticancer effects (Dimitrova et al., 2013; Du et al., 2017).

- **Torularhodin**

Torularhodin ($C_{40}H_{52}O_2$, 564.84 g/mol) resembles torulene structurally. The presence of a carboxyl group at the end of the polyene chain is the only distinction. This substance falls under the category of xanthophylls. According to Di Mascio et al. (1991), torularhodin has a polar appearance and is dark pink in color. *R. mucilaginosa* and *Sporobolomyces ruberrimus* are the principal producers of this substance (Ungureanu et al., 2012; Kanno et al., 2020). According to Table 1, reviews and reports on approximately 70 species belonging to 40 genera from all the fungal groups producing carotenoids have been reported (Valadon, 1976; Johnson and Schroeder, 1995; Yurkov et al., 2008; Barredo, 2012; Manimala and Murugesan, 2017; Kot et al., 2018).

**Fungal carotenoids biosynthetic pathway**

According to Erasun and Johnson (2018), fungi use the mevalonate biosynthetic pathway to produce a large number of crucial secondary metabolites. Among the four large families of compounds that carotenoids serve as a source are: (1) Retinoid compounds or vitamin A Retinal, retinoic acid, and retinol are components of $C_{20}$. (2) Apo-carotenoids $C_{40}$ comprise apocarotenal, bixin, crocetin, food orange 7 color, ionones, and peridinin in addition to abscisic acid. (3) More xanthophyll and carotenoids. The first step in the manufacture of carotenoids in fungal cells is the conversion of acetyl-CoA, which is produced during the fatty acid β-oxidation process in the mitochondria (Figure 2). According to the mevalonic acid pathway, many metabolic processes are catalysed by certain reductases, kinases, and decarboxylases to create isopentenyl pyrophosphate (IPP), a precursor to five-carbon carotenoids. A compound with 20 carbon atoms per molecule, geranyl-geranyl pyrophosphate (GGPP), is created by the addition reactions of three IPPs. Phytoene (C40) is created during the condensation of the two GGPP particles, which is catalysed by phytoene synthase. It is a step in the biosynthesis of lycopene. Lycopene can then be converted into β-carotene, α-carotene, torulene, lutein, torularhodin, zeaxanthin, and astaxanthin, depending on the type of microorganisms (Mussagy et al., 2019).

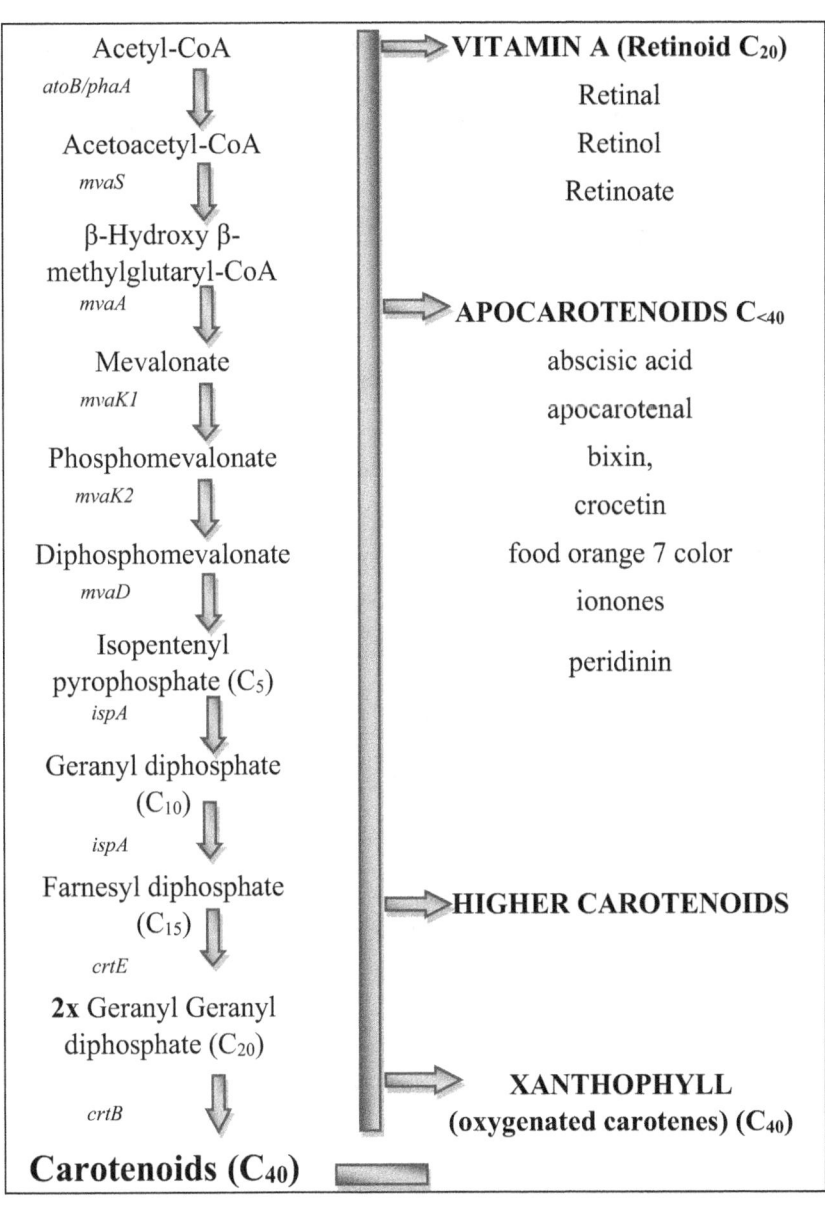

**Figure 2** Carotenoids are made using mevalonate biosynthetic pathways. (Gene and encoded enzymes: atoB/phaA, acetoacetyl-CoA synthase; mvaS, HMG-CoA synthase; mvaA, HMG-CoA

reductase; mvaK1, mevalonate kinase; mvaK2, phosphomevalonate kinase; mvaD, mevalonate-5-diphosphate decarboxy; ispA, FPP synthase;crtE, GGPP synthase; crtB, phytoene synthase).

**Table 1** Fungal Carotenoids.

| Division | Subdivision | Class | Order | Strains | Carotenoids |
|---|---|---|---|---|---|
| MYXOMYCOTA (prokaryotes) | | Myxomycetous fungi (slime molds) | | Many genera and species | β-Carotene, γ-Carotene & lycopene |
| EUMYCOTA | Mastigomycotina | Chytridiomycetes | | | β-Carotene, γ-Carotene, 3,4-didehydrolycopene & neurosporaxanthin |
| EUMYCOTA | Zygomycotina | | Mucorales | Blakeslea trispora (Rhizomucor miehei) | β-carotene, δ-carotene & lycopene |
| | | | | Phycomyces blakesleeanus | β-carotene and δ-carotene |
| | | | | Phycomyces blakesleeanus | β-carotene, δ-carotene, lycopene & phytoene |
| | | | | Choanephora cucurbitarum, Mucor circinelloides | β-carotene |

| Subdivision | Class | Order | Species | Pigment |
|---|---|---|---|---|
| | | | Erwinia uredovora | Lycopene |
| Ascomycotina | Zygomycetes | | Many genera and species | Canthaxanthin, β-cryptoxanthin, echinone, zeaxanthin |
| | Hemiascomycetes | | Un valid | Un valid |
| | Dothideomycetes | | Cercospora nicotianae | Lycopene |
| | Plectomycetes | | Many genera and species | β-Carotene, γ-Carotene & lycopene |
| | Eurotiomycetes | Eurotiales | Paecilomyces | Carotenoids |
| | | | Penicillium atrovenetum | Carotenoids |
| | | | Penicillium herquei | Carotenoids |
| | | | Penicillium species | Tangeraxanthin and 4-ketonostoxanthin |
| | | | Monascus purpureus | Carotenoids |
| | | | Monascus roseus | Canthaxanthin |
| | Pyrenomycetes | | Many genera and species | β-carotene, γ-carotene, 3,4-didehydrolycopene, torulene, aleuriaxanthin, |

| Pezizomycotina | | | | |
|---|---|---|---|---|
| Sordariomycetes | | Discomycetes | | plectaniaxanthin, phillipsiaxanthin |
| | Sordariomycetes | | *Cordyceps* | Carotenoids |
| | | | | β-carotene, γ-carotene, neurosporene, torulene, aleuriaxanthin, plectaniaxanthin & phillipsiaxanthin |
| | Hypocerales | *Fusarium sporotrichioides* | | β-carotene, δ-carotene & lycopene |
| | | *Fusarium moniliform (Gabriella fujikuroi)* | | Phytoene |
| *Neurospora crassa* | | | | β-carotene, δ-carotene & phytoene |

| | | |
|---|---|---|
| Pezizales | *Helvetia crispa, Helvetia esculenta, Helvetia gigas, Helvetia infula, Helvetia lacunosa, Helvetia monachella, Helvetia pallescens, Helvetia pezizoides, Discina ancilis, Discina reticulata, Disciotis venosa, Leptopodia elastica, Paxina acetabulum, Rhizina inflate, Morchella conica, Morchella esculenta, Verpa digitaliformis* | neurosporene, lycopene, β-carotene, γ-carotene, α-carotene, flavoxanthin, mutatochrome, β-zeacarotene, aleuriaxanthin, aleuriaxanthin ester, Astaxanthin, canthaxanthin, β-cryptoxanthin, lycoxanthin, neurosporaxanthin, rubixanthin, plectaniaxanthin, torularhodin, 3,4-dehydro lycopene, hydroxy-α-carotene, dihydroxy-C-carotene, 1,2,1',2'-tetrahydro-1, 1'-dihydroxy lycopene |

| | | | | |
|---|---|---|---|---|
| **Basidiomycotina** | | | Saccharomycetes | *Yarrowia, Zygosaccharomyces rouxii* (yeast) | Carotenoids |
| | | | | *Cyberlindnera jadinii (Candida utilis)* GE, *Pichia pastoris* by GE, *Saccharomyces cerevisiae* by GE | Non carotenogensis yeast converted by genetic engineering to carotenoids producers on large scales |
| | Loculoascomycetes | Filobasidiales | | *Dioszegia hungarica* | Carotenoids |
| | | Tremellales | | *Cryptococcus aerius, laurentii, magnus & victoria* | Carotenoids |
| | | Cystofilobasidiales | | *Cystofilobasidium capitatum* | β-carotene, γ-carotene, torulene & torularhodine |
| Hymenomycetes | | Leucosporidiales | | *Leucosporidiella muscorum* | β-carotene, γ-carotene, astaxanthin, canthaxanthin, cryptoxanthin & diadinoxanthin |

115

| | | | |
|---|---|---|---|
| Gasteromycetes | | | β-carotene, γ-carotene |
| Teliomycetes | | | β-carotene, γ-carotene, torulene, phytoene, cryptoxanthin |
| Microbotryomycetes, | Sporidiobolales | Rhodotorula aurantiaca, Rhodotorula acheniorum, Rhodotorula graminis, Rhodotorula glutinis, Rhodotorula minuta & Rhodotorula mucilaginosa (Red yeasts) | Carotenoids β-carotene, γ-carotene, torulene, torularhodin, antheraxanthin & echinenone |
| | | Rhodosporidium babjevae, Rhodosporidium Minuta, Rhodosporidium diobovatum & Rhodosporidium sphaerocarpum | β-carotene, torulene & torularhodine |

| Agaricomycotina | | |
|---|---|---|
| Tremellomycetes | | Sporidiobolus salmonicolor, Sporobolomyces pararoseus, Sporobolomyces roseus, Sporobolomyces salmonicolor |
| | Xanthophyllomyces dendrorhous Phaffia rhodozyma (Rhodomyces dendrorhous) & Phaffia paxilli east | Astaxanthin |
| Agaricomycetes Edible mushrooms | Russula virescens Cantharellus cibarius, Cantharellus minor, Cantharellus friesii & Cantharellus cinnabarinus | β-carotene & lycopene Canthaxanthin |

| | | |
|---|---|---|
| **Deuteromycot ina** | | |
| | | |
| | | |
| | *Choanephora cucurbitarum* Liakopoulou-Kyriakides | β-carotene, γ-carotene & tonularhodine |

## Carotenoids located within the mould cells

The structural and functional components known as fungal carotenoids are produced in the cytoplasm, stored in cytoplasmic vesicles (lipid globules), and transported to the fungal plasma membrane portion. In fungal cells, carotenoids play crucial roles as antioxidants against free radicals, protective molecules or photoreceptors against harmful or intense light and radiation, and maintainers of membrane fluidity and stability (Figure 3). In order to complete the fungal life cycle, they also serve as a source of the sex hormones (Erasun and Johnson, 2018).

For instance, *Blakeslea trispora* and *Mucor mucedo* contain β-carotene, according to Sahadevan et al. (2013), where, male and female sex hormones are produced by zygomycetes, which aid in bringing together and contacting male and female gametes in order to complete sexual reproduction and the fungal life cycle.

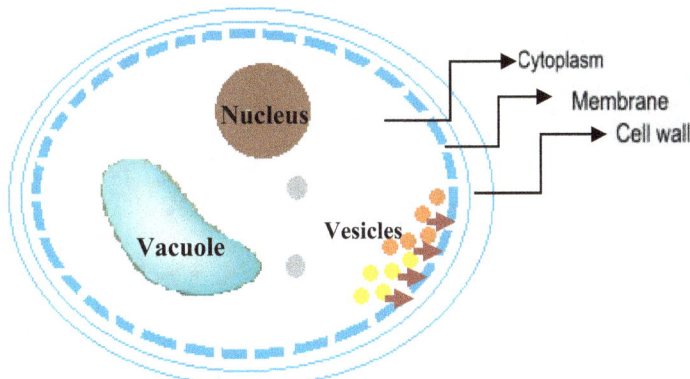

**Figure 3** A diagram demonstrated the carotenoids transferred from storage in cytoplasmic vesicles to the fungal plasma membrane fraction.

**Fungal carotenoids therapy**

Numerous writers (Eman et al., 2018, Tan and Norhaizan, 2019, Ramesh et al., 2019) reported on the bioactivity of carotenoids (Figure 4). Carotenoids are well-known to be essential for maintaining good health in human. They can avoid a lack of vitamin A, which is considered to be crucial for promoting growth, embryonic development, and vision. Carotenoids are abundant in membranes and other lipophilic compartments, which are determined by their lipophilicity (Stahl and Sies, 2015). Additionally, according to Seel et al. (2020), polar carotenoids can control membrane fluidity.

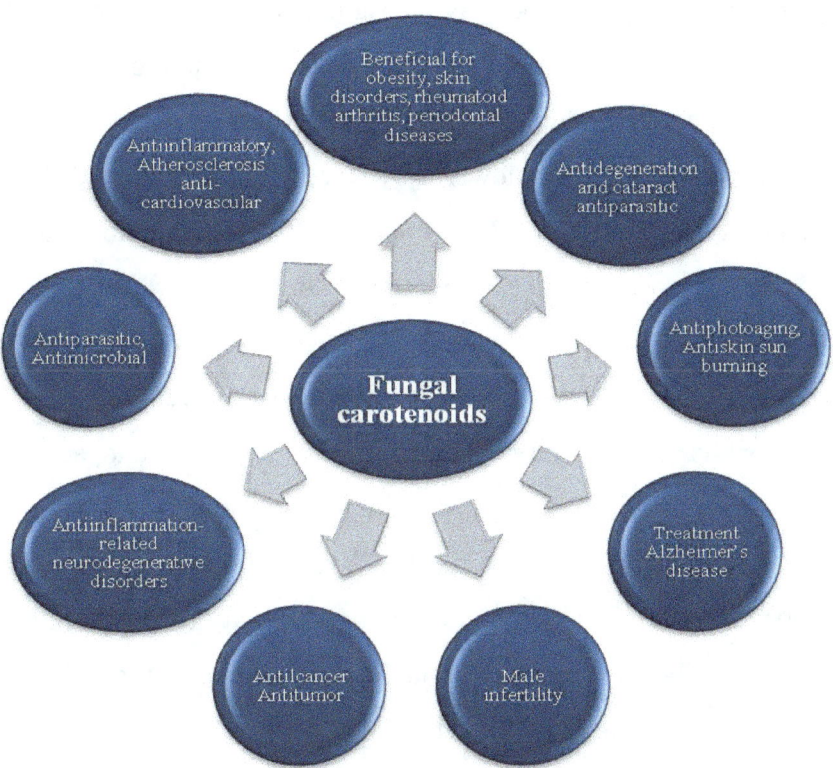

**Figure 4** Bioactivities of fungal carotenoids for human health.

By preventing cataracts and age-related macular degeneration, lutein and zeaxanthin carotenoids are essential for the protection of the retina (Eggersdorfer and Wyss, 2018). Torularhodin is a carotenoid that is mostly generated by *Sporobolomyces* and *Rhodotorula* species; it has potent antibacterial capabilities and may develop into a new type of natural antibiotic (Kot et al., 2018). According to McEneny et al. (2013), the effectiveness of lycopene use is to demonstrate an anti-inflammatory property and may be trigger an organism's immune response, which lowers the chance of developing atherosclerosis and other cardiovascular illnesses. Additionally, systolic blood pressure was dramatically lowered by lycopene (Kim et al., 2011). According to research by Ramesh et al. (2019), astaxanthin can prevent atherosclerotic cardiovascular illnesses by reducing oxidative stress and inflammation. Lycopene and β-carotene have been linked to bone health, osteoporosis prevention, and osteoporosis risk reduction, particularly in postmenopausal women (Rao and Rao (2007). The effects of β-cryptoxanthin were also discussed (Eggersdorfer and Wyss, 2018). Consuming lycopene has been shown to enhance bone density by decreasing bone resorption and to protect against type 2 diabetes by promoting glucose homeostasis (Walallawita et al., 2020).

Additionally, using lycopene to treat male infertility improved sperm motility, sperm motility index, sperm morphology, and functional sperm concentration, which ultimately led to a 36% rise in successful pregnancies (Rao and Rao, 2007). According to Eggersdorfer and Wyss (2018) and Bonet et al. (2020), skin conditions, rheumatoid arthritis, and periodontal illnesses, carotenoids may help with weight control and obesity. Lycopene's beneficial effects in the management of Alzheimer's disease were also investigated (McEneny et al., 2013). Due to its potent antioxidant action, torularhodin can be employed as a neuroprotective drug against $H_2O_2$-induced oxidative stress, while lutein is assumed to be connected to the potential management of inflammation-related neurodegenerative illnesses (Wu et al., 2015).

A red pigment that accumulated in *S. cerevisiae* mutants has an interesting relationship to abnormal protein aggregation and the development of amyloid fibrils. This pigment can bind to amyloid fibrils and interfere with their interactions with chaperones, which in turn prevent prion "multiplication" and the formation of amyloid fibrils (Nevzglyadova et al., 2011). As a blue light filter, β-carotene can be utilized for sun protection, the avoidance of sunburn, defense against photo-oxidative damage to lipids, proteins, and DNA, and the prevention of skin cancer and premature ageing of the skin (Sies and Stahl, 2004). According to Ajila and Brar (2012), β-carotene suppresses neoplasm promotion and progression as well as lowers the risk of getting neoplastic disorders. In addition to hepatoma, leukaemia, uveal melanoma, and other malignancies, the anticancer activity of various carotenoids, such as α-carotene, β-carotene, lycopene, torulene, and torularhodin, has been examined (Du et al., 2016).

**Fungal carotenoids industry**

On big industrial scales, fungus carotenoids are highly valued and their popularity is growing among consumers. They are desired for nutritional food colorants in cakes, confectionaries, candies, pudding, jelly, fruits, breakfast cereals, pasta, sauces, processed cheese, fruit, vitamin-enriched milk products, energy drinks and beverages, cosmetics and perfume, dyeing of wood, texture, leather, papers and painting, and supplemental foods (Narsing-Rao et al., 2017; Meléndez-Mart). They are also desired for use as natural colourants in chemical and folk and modern medicine. Although *Mucor circinelloides*, *Phycomyces blakesleeanus*, and *B. trispora* are widely recognized -carotene producers, where, *B. trispora* is the predominant nonpathogenic and nontoxigenic species employed in industrial production (Böhme et al., 2006). *Rhodotorula* deserves special attention for increasing the production of carotenoids through various stress factors, where *R. glutinis*, *R. mucilaginosa*, and *Sporidiobolus salmonicolor* produce significantly higher amounts of carotenoids for industrial purposes in the case of oxidative, osmotic, and salt stress (Mannazzu et al., 2015). It was discovered that *Rhodotorula toruloides* produced more carotenoids when the temperature was low in media containing agro-industrial waste, potato wastewater, and glycerol. Low temperature and the introduction of osmotic stress accelerated the manufacture of β-carotene, accounting for up to 73.9% of the total carotenoid concentration. While white-light irradiation enhanced torularhodin production (up to 20.0%), oxidative stress increased the efficiency of torulene synthesis by yeast (up to 82.2% vs. other circumstances) (Kot et al., 2019).

Economic efficiency allows for a large rise in the biotechnological production of carotenoids when expenses are reduced by using waste or by-products from other biotechnologies as the primary substrate for fungi (Da Silva et al., 2020). Interesting research revealed that discarded brewer's yeast *Rhodotorula* strains may also be used to create carotenoids in addition to the raw glycerin from the manufacturing of biodiesel (Rodrigues et al., 2019). *Sporobolomyces roseus*, *R. glutinis*, *R. mucilaginosa*, and *Cystofilobasidium capitatum* were shown to produce carotenoids from used effectively coffee grounds (Obruca et al., 2015). Incubated in media containing parboiled rice water, crude glycerol, and sugar cane molasses, three yeast strains from the species *Sporidiobolus pararoseus*, *R. mucilaginosa*, and *Pichia fermentans* isolated in Brazilian forests were discovered to produce cryptoxanthin and β-carotene (Pereira et al., 2019). *R. mucilaginosa* may produce β-carotene in culture using fruit wastes from orange, pomegranate, and pineapple (Korumilli and Mishra, 2014).

It is obvious that different approaches must be taken when considering the large-scale biotechnological production of different carotenoids; however, it is important to keep in mind that the production of carotenoids by fungi depends on a variety of variables. The presence of some compounds (such as ethanol and acetic acid) in the growth media, as well as carbon and nitrogen sources, light, temperature, aeration, metal ions, and especially some trace elements, are among these parameters. It was demonstrated that the synthesis of pigments could increase with the usage of carbon sources like ethanol (Chreptowicz et al., 2019). White light has a good impact on their production. Another significant element that affects the production of carotenoids is temperature (Zhang et al., 2014). According to research on *R. glutinis*, β carotene and torulene synthesis were favored at a temperature of 25 °C, but torularhodin biosynthesis favored at a temperature range of 30–35 °C (Malisorn et al., 2008).

Another crucial element affecting carotenoid production is proper aeration (Saenge et al., 2011). The amount of β-carotene produced by the *S. cerevisiae* strain increased by two times when 0.5 M NaCl was added to the fresh water used to make the nutritional media. For the synthesis of carotenoids by different species of *Rhodotorula*, metal ions (including Ba, Fe, Mg, Ca, Zn, and Co) and particularly some trace elements (like Al, Zn, and Mn) are crucial (Guo et al., 2019). The greatest titer of recombinant β-carotene produced to date was reported and discussed the potential of the ascomycetous yeast species *Y. lipolytica* as a -carotene-producing cell factory (Larroude et al., 2017).

**Metabolic engineering**

Metabolic and genetic engineering is the best method to boost carotenoids production and lower production costs in yeast. Application of genetic engineering in yeasts through the following steps: (1) selection of high-yielding strains of carotenoids producers; (2) enhancement of their productivity through metabolic engineering; (3) acquisition of carotenogenic genes from high-yielding yeast strains like *Erwinia uredovora*, *Agrobacterium aurantiacum*, and *Xanthophyllomyces dendrorhus*; and (4) Inserting carotenogenic genes (β-carotene, lycopene, and astaxanthin gene) into non-carotenogenic yeast like *S. cerevisiae*, *P. pastoris* and *C. utilis* (Erasun and Johnson, 2018). Several marine yeast species that have been isolated produce carotenoids. Astaxanthin has been obtained, in particular, from the genera *Xanthophyllomyces*, *Rhodotorula*, and *Phaffia* (Britton, 2020).

Carotenoids are produced even by protists-resembling fungi, including thraustochytrids. Planktonic and benthonic forms inhabit a variety of habitats including sediments of mangroves, estuaries, and deep-sea ecosystems, and they have a wide geographic distribution from the polar to tropical regions (Rodrigues et al., 2019). According to Mapelli-Brahm et al. (2020), *Thraustochytrium* strains ONC-T18 and CHN-1, *Thraustochytriidae* sp. AS4-A1, and *Aurantiochytrium* sp. KH105 produce a variety of carotenoids, including β-carotene, astaxanthin, zeaxanthin, cantaxant Engineering methods have made it possible to produce more carotenoids, sometimes with astaxanthin content that is nine times higher, as in the instance of *Aurantiochytrium* sp. SK4 (**Petrik et al., 2014**). To increase our understanding of the sources of carotenoids for application in the biotechnological area, it is essential to develop genetic methods and genome sequencing applied to thraustochytrids.

According to research by Tramontin et al. (2019), it has been demonstrated that recombinant microbial cell factories may be created on the basis of the oleaginous yeast *Yarrowia lipolytica* to generate astaxanthin through submerged fermentation. The capability of a yeast-based method for β-carotene production was proven by a study focused on metabolic engineering of *S. cerevisiae* (Fathi et al., 2021). According to a report, chemical mutagenesis developed by *B. trispora* variants that produced 100 times more β-carotene than the wild-type strain. According to Li et al. (2020), the same organism is also suggested for the commercial synthesis of lycopene.

### Conclusion

Biotechnological platforms for industrial and medicinal production are interested in fungal carotenoids. Fungal carotenoids that occur naturally have been approved for human consumption, and biologically produced antioxidant activity is said to be higher than that of chemically created counterparts. In order to meet the demand for industrial production, this chapter describes the most recent research being done to genetically modify carotenogenic and non-carotenogenic fungi to enhance carotenoid production in model chassis. They also highlighted a few commercially significant fungal carotenoids. This chapter also provided an overview of the great potential for using mutation analysis and functionality investigations to genetically design the unique structures and biosynthesis processes of fungal carotenoids. Nevertheless, very little is known about the output of carotenoids from fungal sources for industrialization. Consequently, in order to identify and facilitate fungal carotenogenesis, it is necessary to screen more robust and prolific candidates, modify the growing conditions, and employ highly effective genetic modification tools and procedures.

**References**

Aasen, I.M.; Ertesvåg, H.; Heggeset, T.M.B.; Liu, B.; Brautaset, T.; Vadstein, O.; Ellingsen, T.E. Thraustochytrids as production organisms for docosahexaenoic acid (DHA), squalene, and carotenoids. Appl. Microbiol. Biotechnol. 2016, 100, 4309–4321.

Ahmad F, Fanning K, Netzel M, Turner W, Li Y, Schenk PM (2014). Profiling of carotenoids and antioxidant capacity of microalgae from subtropical coastal and brackish waters. Food Chemistry 165:300-306.

Ajila, C.; Brar, S. Role of dietary antioxidants in cancer. In Nutrition, Diet and Cancer; Shankar, S., Srivastava, R.K., Eds.; Springer: Dordrecht, Germany, 2012; pp. 377–412.

Ambati, R.R.; Phang, S.M.; Ravi, S.; Aswathanarayana, R.G. Astaxanthin: Sources, extraction, stability, biological activities and its commercial applications—A review. Mar. Drugs 2014, 12, 128–152.

Barredo JL (2012). Microbial carotenoids from fungi: Methods and protocols, methods in molecular biology, Springer Science+Business Media New York.

Bhosale, P.; Gadre, R.V. Manipulation of temperature and illumination conditions for enhanced beta-carotene production by mutant 32 of Rhodotorula glutinis. Lett. Appl. Microbiol. **2002**, 34, 349–353.

Black, R.E.; Allen, L.H.; Bhutta, Z.A.; Caulfield, L.E.; De Onis, M.; Ezzati, M.; Mathers, C.; Rivera, J. Maternal and child undernutrition: Global and regional exposures and health consequences. Lancet **2008**, 371, 243–260.

Böhme, K.; Richter, C.; Pätz, R. New insights into mechanisms of growth and _-carotene production in Blakeslea trispora. Biotechnol. J. **2006**, 1, 1080–1084.

Bonet, M.L.; Ribot, J.; Galmés, S.; Francisca Serra, F.; Palou, A. Carotenoids and carotenoid conversion products in adipose tissue biology and obesity: Pre-clinical and human studies.

Biochim. Biophys. Acta Mol. Cell Biol. Lipids **2020**, 1865, 158676

Britton, G. Carotenoid research: History and new perspectives for chemistry in biological systems. Biochim. Biophys. Acta Mol. Cell Biol. Lipids **2020**, 1865, 158699.

Chreptowicz, K.; Mierzejewska, J.; Tkácová, J.; Młynek, M.; Certik, M. Carotenoid-producing yeasts: Identification and characteristics of environmental isolates with a valuable extracellular enzymatic activity. Microorganisms **2019**, 7, 653.

Da Silva, S.R.S.; Stamford, T.C.M.; Albuquerque,W.W.C.; Vidal, E.E.; Stamford, T.L.M. Reutilization of residual glycerin for the produce _-carotene by Rhodotorula minuta. Biotechnol. Lett. **2020**, 42, 437–443.

Di Mascio, P.; Murphy, M.E.; Sies, H. Antioxidant defense systems: The role of carotenoids, tocopherols, and thiols. Am. J. Clin. Nutr. **1991**, 53, 194S–200S.

Dimitrova, S.; Pavlova, K.; Lukanov, L.; Korotkova, E.; Petrova, E.; Zagorchev, P.; Kuncheva, M. Production of metabolites with antioxidant and emulsifying properties by Antarctic strain Sporobolomyces salmonicolor AL1. Appl. Biochem. Biotechnol. **2013**, 169, 301–311.

Du, C.; Guo, Y.; Cheng, Y.; Han, M.; Zhang, W.; Qian, H. Anti-cancer effects of torulene, isolated from Sporidiobolus pararoseus, on human prostate cancer LNCaP and PC-3 cells via a mitochondrial signal pathway and the down-regulation of AR expression. RSC Adv. **2017**, 7, 2466–2474.

Du, C.; Li, Y.; Guo, Y.; Han, M.; Zhang,W.; Qian, H. The suppression of torulene and torularhodin treatment on the growth of C-3 xenograft prostate tumors. Biochem. Biophys. Res. Commun. **2016**, 469, 1146–1152.

Eggersdorfer, M.;Wyss, A. Carotenoids in human nutrition and health. Arch. Biochem. Biophys. **2018**, 652, 18–26.

Eman Mostafa M, Khallaf IS, Nassar SM (2018). Antibacterial activities of some yeast strains and GC/MS analysis of *Rhodotorula mucilaginosa* AUMC13565 bioactive metabolites. Assiut University Journal of Botany and Microbiology 47(2):55-70.

Erasun EC, Johnson EA (2018). Fungal carotenoids. Chapter in Applied Mycology and Biotechnology pp1-55.

Fathi, Z.; Tramontin, L.R.R.; Ebrahimipour, G.; Borodina, I.; Darvishi, F. Metabolic engineering of Saccharomyces cerevisiae for production of _-carotene from hydrophobic substrates. FEMS Yeast Res. **2021**, foaa068.

Grune, T.; Lietz, G.; Palou, A.; Ross, A.C.; Stahl, W.; Tang, G.; Thurnham, D.; Yin, S.A.; Biesalski, H.K. Beta-carotene is an important vitamin A source for humans. J. Nutr. **2010**, 140, 2268S–2285S.

Gul, K.; Tak, A.; Singh, A.K.; Singh, P.; Yousuf, B.; Wani, A.A.; Yildiz, F. Chemistry, encapsulation, and health benefits of β-carotene, A review.Cogent Food Agric. **2015**, 1, 1018696.

Guo, Y.; Xie, S.; Yuan, J.S.; Kao, K.C. Effects of seawater on carotenoid production and lipid content of engineered Saccharomyces cerevisiae. Fermentation **2019**, 5, 6.

Hausmann, A.; Sandmann, G. A single five-step desaturase is involved in the carotenoid biosynthesis pathway to _-carotene and torulene in Neurospora crassa. Fung. Gen. Biol. **2000**, 30, 147–153.

Jannel, S.; Caro, Y.; Bermudes, M.; Petit, T. Novel insights into the biotechnological production of Haematococcus pluvialis-derived astaxanthin: Advances and key challenges to allow its industrial use as novel food ingredient. J. Mar. Sci. Eng. **2020**, 8, 789.

Johnson EA, Schroeder W (1995). Microbial carotenoids. Advance Biochemistry Engineering Biotechnology 53:119-178.

Kanno, K.Y.F.; Karp, S.G.; Rodrigues, C.; Tanobe, V.O.A.; Soccol, C.R.; da Costa Cardoso, L.A. Influence of organic solvents in the extraction and purification of torularhodin from Sporobolomyces ruberrimus. Biotechnol. Lett. **2020**.

Kim, J.Y.; Paik, J.K.; Kim, O.Y.; Park, H.V.; Lee, J.H.; Jang, Y.; Lee, J.H. Effects of lycopene supplementation on oxidative stress and markers of endothelial function in healthy men. Atherosclerosis **2011**, 215, 189–195.

Kirti K, Amita S, Priti S, Kumar AM, Jyoti S (2014). Colorful world of microbes: carotenoids and their applications. Review Article. Advances in Biology 13:1-13.

Korumilli, T.; Mishra, S. Carotenoid production by Rhodotorula sp. on fruit waste extract as a sole carbon source and optimization of key parameters. Iran J. Chem. Chem. Eng. **2014**, 33, 89–99.

Kot AM, Błażejak S, Gientka I, Kieliszek M, Bryś J (2018). Torulene and torularhodin: new fungal carotenoids for industry. Microbial Cell Factories 17(49):3-14.

Kot, A.M.; Bła ̇zejak, S.; Gientka, I.; Kieliszek, M.; Brys, J.; Reczek, L.; Pobiega, K. Effect of exogenous stress factors on the biosynthesis of carotenoids and lipids by Rhodotorula yeast strains in media containing agro-industrial waste.WorldJ. Microbiol. Biotechnol. **2019**,35,157.

Kuczynska P, Jemiola-Rzeminska M (2017). Isolation and purification of all-trans diadinoxanthin and all-trans diastaxanthin from diatom *Phaeodactylum tricornutum*. Journal Applied Phycology 29:79-87.

Larroude, M.; Celinska, E.; Back, A.; Thomas, S.; Nicaud, J.M.; Ledesma-Amaro, R. A synthetic biology approach to transform Yarrowia lipolytica into a competitive biotechnological producer of b-carotene. Biotechnol. Bioeng. **2017**, 115, 464–472.

Li, L.; Liu, Z.; Jiang, H.; Mao, X. Biotechnological production of lycopene by microorganisms. Appl. Microbiol. Biotechnol. **2020**, 104, 10307–10324

Malik K, Tokkas J, Goyal S (2012). Microbial pigments. Review Article. International Journal of Microbial Resource Technology 1(4):361-365

Malisorn, C.; Suntornsuk, W. Optimization of _-carotene production by Rhodotorula glutinis DM28 in fermented radish brine. Bioresour. Technol. **2008**, 99, 2281–2287.

Manimala MR, Murugesan R (2017). Carotenoid pigment production from yeast: Health benefits and their industrial applications. International Journal of Chemical Studies 5(6):392-395.

Mannazzu, I.; Landolfo, S.; Lopes da Silva, T.; Buzzini, P. Red yeasts and carotenoid production: Outlining a future for non-conventional yeasts of biotechnological interest. World J. Microbiol. Biotechnol. **2015**, 31, 1665–1673.

Manowattana, A.; Techapun, C.; Laokuldilok, T.; Phimolsiripol, Y.; Chaiyaso, T. Enhancement of β-carotene-rich carotenoid production by a mutant Sporidiobolus pararoseus and stabilization of its antioxidant activity by microencapsulation. J. Food Process. Preserv. **2020**, 44, e14596

Mapelli-Brahm, P.; Barba, F.J.; Remize, F.; Garcia, C.; Fessard, A.; Khaneghah, A.M.; Sant'Ana, A.S.; Lorenzoe, J.M.; Montesano, D.; Meléndez-Martínez, A.J. The impact of fermentation processes on the production, retention and bioavailability of carotenoids: An overview. Trends Food Sci. Technol. **2020**, 99, 389–401. [

Mata-Gómez LC, Montañez JC, Méndez-Zavala A, Aguilar CN (2014). Biotechnological production of carotenoids by yeasts: An overview Review. Microbial Cell Factories 13(12):1-.

McEneny, J.;Wade, L.; Young, I.S.; Masson, L.; Duthie, G.; McGinty, A.; McMaster, C.; Thies, F. Lycopene intervention reduces inflammation and improves HDL functionality in moderately overweight middle-aged individuals. J. Nutr. Biochem. **2013**, 24,163–168.

Meléndez-Martínez AJ, Stinco CM, Mapelli-Brahm P (2019). Skin carotenoids in public health and nutri-cosmetics: The emerging roles and applications of the UV radiation-absorbing colorless carotenoids phytoene and phytofluene. Review. Nutrients 11:1-41.

Metlicar, V.; Vovk, I.; Albreht, A. Japanese and Bohemian Knotweeds as Sustainable Sources of Carotenoids. Plants **2019**, 8, 384.

Mussagy, C.U.; Winterburn, J.; Santos-Ebinuma, V.C.; Pereira, J.F.B. Production and extraction of carotenoids produced by microorganisms. Appl. Microbiol. Biotechnol. **2019**,103,1095–1114.

Nagal S. and Panda I.P. (2022) Isolation, identification and characterization of β- carotene producing endophytic fungi- *Penicillium citrinum* from bark of Taxus baccata

Narsing-Rao MP, Xiao M, Li WJ (2017). Fungal and bacterial pigments: secondary metabolites with wide applications. Frontiers in Microbiology 8(1113):1-13.

Nevzglyadova, O.V.; Kuznetsova, I.M.; Mikhailova, E.V.; Artamonova, T.O.; Artemov, A.V.; Mittenberg, A.G.; Kostyleva, E.I.; Turoverov, K.K.; Khodorkovskii, M.A.; Soidla, T.R. The effect of red pigment on amyloidization of yeast proteins. Yeast **2011**, 28; 505–526.

Obruca,S.;Benesova,P;Kucera,D.;Petrik,S.;Marova,I.Biotechnological conversion of spent coffee grounds into polyhydroxyalkanoates and carotenoids.New Biotechnol. **2015**,32,569–574.

Pereira, R.N.; da Silveira, J.M.; de Medeiros Burkert, J.F.; Ores, J.d.C.; Burkert, C.A.V. Simultaneous lipid and carotenoid production by stepwise fedpbatch cultivation of Rhodotorula mucilaginosa with crude glycerol. Braz. J. Chem. Eng. **2019**, 36, 1099–1108.

Petrik, S.; Obruca, S.; Benesova, P.; Marova, I. Bioconversion of spent coffee grounds into carotenoids and other valuable metabolites by selected red yeast strains. Biochem. Eng. J. **2014**, 90, 307–315.

Raghukumar, S. Ecology of the marine protists, the Labyrinthulomycetes (Thraustochytrids and Labyrinthulids). Eur. J. Protistol. 2002, 38, 127–145.

Ramesh C, Vinithkumar NV, Kirubagaran R, Venil CK, Dufossé L (2019). Multifaceted applications of microbial pigments: current knowledge, challenges and future directions for public health implications. Review Microorganisms 7(7):186.

Rao, A.V.; Rao, L.G. Carotenoids and human health. Pharmacol. Res. **2007**, 55, 207–216.

Rodrigues, T.A.; Schueler, T.A.; da Silva, A.J.R.; Sérvulo, E.F.C.; Oliveira, F.J.S. Valorization of solid wastes from the brewery and biodiesel industries for the bioproduction of natural dyes. Braz. J. Chem. Eng. **2019**, 36, 99–107

Rodríguez-Sáiz, M.; Fuente, J.; Barredo, J. Xanthophyllomyces dendrorhous for the industrial production of astaxanthin. Appl. Microbiol. Biotechnol. **2010**, 88, 645–658.

Saenge, C.; Cheirsilp, B.; Suksaroge, T.T.; Bourtoom, T. Potential use of oleaginous red yeast Rhodotorula glutinis for the bioconversion of crude glycerol from biodiesel plant to lipids and carotenoids. Process. Biochem. **2011**, 46, 210–218.

Sahadevan Y, Richter-Fecken M, Kaerger K, Volgt K, Boland W (2013). Early and late trisporoids differentially regulate β-carotene production and gene transcript levels in the mucoralean fungi *Blakeslea trispora* and *Mucor mucedo*. Applied and Environmental Microbiology 29(23):7466-7475.

Seel, W.; Baust, D.; Sons, D.; Albers, M.; Etzbach, L.; Fuss, J.; Lipski, A. Carotenoids are used as regulators for membrane fluidity by Staphylococcus xylosus. Sci. Rep. **2020**, 10, 330.

Sharma, R.; Ghoshal, G. Optimization of carotenoids production by Rhodotorula mucilaginosa (MTCC-1403) using agro-industrial waste in bioreactor: A statistical approach. Biotechnol. Rep. **2020**, 25, e00407.

Sies, H.; Stahl, W. Nutritional protection against skin damage from sunlight. Annu. Rev. Nutr. **2004**, 24, 173–200.

Stahl, W.; Sies, H. Bioactivity and protective effects of natural carotenoids. Biochim. Biophys. Acta **2005**, 1740, 101–107.

Suen, Y.L.; Tang, H.; Huang, J.; Chen, F. Enhanced production of fatty acids and astaxanthin in Aurantiochytrium sp. by the expression of Vitreoscilla hemoglobin. J. Agric. Food Chem. 2014, 62, 12392–12398.

Tan BL, Norhaizan EM (2019). Carotenoids. How effective are they to prevent age-related diseases. Review Molecules 1801:1-23.

Tramontin, L.R.R.; Kildegaard, K.R.; Sudarsan, S.; Borodina, I. Enhancement of astaxanthin biosynthesis in oleaginous yeast

Yarrowia lipolytica via microalgal pathway. Microorganisms **2019**, 7, 472.

Tuli HS, Chaudhary P, Beniwal V, Sharma AK (2015). Microbial pigments as natural color sources: current trends and future perspectives. Review Journal Food Science Technology 52(8):4669-4678.

Ungureanu, C.; Ferdes, M.; Chirvase, A.A. Torularhodin biosynthesis andn extraction by yeast cells of Rhodotorula rubra. Rev. Chim. **2012**, 63, 316–318.

Valadon L. R. G. Carotenoids As Additional Taxonomic Characters In Fungi: A Review. Trans. Br. mycol. Soc. 67 (1) 1-15 (1976)

Venil CK, Zakaria ZA, Ahmad WA (2013). Bacterial pigments and their applications. Review. Process Biochemistry 48:1065-1079.

Walallawita, U.S.; Wolber, F.M.; Ziv-Gal, A.; Kruger, M.C.; Heyes, J.A. Potential role of lycopene in the prevention of postmenopausal bone loss: Evidence from molecular to clinical studies. Int. J. Mol. Sci. **2020**, 21, 7119.

Wang Q, Liu D, Yang Q, Wang P (2017). Enhancing carotenoid production in *Rhodotorula mucilaginosa* KC8 by combining mutation and metabolic engineering. Ann Microbiology 67:425-431.

Wei, C.; Wu, T.; Ao, H.; Qian, X.; Wang, Z.; Sun, J. Increased torulene production by the red yeast, *Sporidiobolus pararoseus*, using citrus juice. Prep. Biochem. Biotechnol. **2019**, 10, 1–8.

Wu, J.L.; Wang, W.Y.;Cheng, Y.L.;Du,C.; Qian, H. Neuroprotective effects of torularhodin against H2O2-induced oxidative injury and apoptosis in PC12 cells. Pharmazie **2015**,70,17–23.

Yurkov AM, Vustin MM, Tyaglov BV, Maksimova IA, Sineokiy SP (2008). Pigmented Basidiomycetous yeasts are a promising source of carotenoids and ubiquinone Q. Microbiology 77(1):1-6.

Zhang, Z.; Zhang, X.; Tan, T. Lipid and carotenoid production by Rhodotorula glutinis under irradiation/high-temperature and dark/low-temperature cultivation. Bioresour. Technol. **2014**, 157, 149–153.

Zhao Y, Guo L, Xia Y, Zhuang X, Weihua Chu W (2019). Isolation, identification of carotenoid-producing *Rhodotorula* sp. from marine environment and optimization for carotenoid production Marin Drugs 17(161):1-9.

# CHAPTER V
# Microalgae-Derived Pigments: An Introduction to Their Biosynthesis and Applications

**Amira Sabry,** *PhD*

Protein Research Department, Genetic Engineering and Biotechnology Research Institute (GEBRI), City of Scientific Research and Technological Applications (SRTA-City), New Borg EL-Arab, 21934, Alexandria, Egypt.

**Abstract**

Recently, natural pigments are facing a fast-growing global market due to the increase of people's awareness of health and the discovery of their novel pharmacological effects.

Microalgae are acknowledged as one of the main photosynthesizers of naturally-derived pigments with bioactive properties and biotechnological applications in areas such as the food, pharmaceutical, cosmeceutical, and nutraceutical industries, owing to their antioxidant, antidiabetic, anti-obesity, anti-inflammatory, antiaging, antimalarial, and neuroprotective properties. Compared with synthetic counterparts, microalgal pigments have no toxic or side effects on the human body and even act as nutritional enhancers with various biological activities.

This chapter provides a comprehensive review of the current knowledge of microalgal pigments classification, distribution, function, and application fields. Furthermore, the most significant factors affecting their production are discussed. Finally, the future perspectives are proposed.

**Keywords:** microalgae; carotenoids; chlorophylls; phycobiliproteins; antioxidant; antidiabetic; anticancer

## 1. Introduction

In recent years, consumer demand for naturally sourced products to promote health and reduce disease has grown steadily. This demand has entailed an increased interest in new natural sources of food, pharmaceutical, and cosmetic products. In this context, the marine environment has been considered a potential reservoir of natural compounds. Among the organisms present in this environment, it is worth highlighting algae.

Algae constitute an extremely diverse polyphyletic group of photosynthetic organisms, which contain multiple species ranging from unicellular microalgae (prokaryotic cyanobacteria) along with eukaryotic photosynthetic microorganisms to multicellular macroalgae (eukaryotic algae) and their length vary between 0.2 μm and 65 cm existing in about 50,000 algal species. Algal species are known to subsist in coastal and aquatic habitats; however, they are also reported in extreme conditions such as hot springs, polar regimes, and salt pans etc. They have the ability to convert solar energy and carbon dioxide into biomass and oxygen (Show et al., 2017; Keykha Akhar et al., 2021; Pereira et al., 2021; Uma et al., 2023).

Microalgae are eukaryotic organisms that have evolved through a series of primary and secondary endosymbiosis. They are ubiquitous in nature and are adapted to cope with a wide range of environmental conditions. Microalgae are classified into different species, the most important of which are green algae (Chlorophyceae), blue-green algae (Cyanophyceae [cyanobacteria]), golden algae (Chrysophyceae), yellow-green algae (Xanthophyceae), and diatom (Bacillariophyceae) (Fu et al., 2019; Keykha Akhar et al., 2021; Sreenikethanam et al., 2022). Microalgae are envisioned as promising candidates in the renewable energy market as third generation biofuels besides being ideal agents for natural supplements, medicines and feeds. As per the global algae market analysis report 2019–2025, algae market is expected to grow by
US$414.8 Thousand. They are an affluent source of bioactive compounds with an outstanding physiological activity, namely proteins, amino acids, pigments, lipids, fatty acids of dietary value, polysaccharides and auxins etc. Recent studies have identified anti-bacterial, antioxidant, anti-inflammatory, antitumor and antiviral properties, in addition to the established phenomenon that algal biomass itself acts as dietary and nutrient source of food and feed standards. However, commercial cultivation of microalgae began in recent decades, and it still needs enormous research attention (Sun et al., 2023; Uma et al., 2023).

## 2. Microalgae-derived pigments

Microalgae represent the important natural pigment sources, which play an irreplaceable role in photosynthetic carbon fixation and cell growth. They can be cultured sustainably, in an eco-friendly manner, and are renewable on an industrial scale without the restriction of seasonal, climatic and environmental conditions. They are also outstanding for their high pigment content, fast growth rate, ability to grow in stress conditions, and the fact that they do not require arable land, thus becoming one of the most promising and competitive sources of natural pigments (Sun et al., 2023).

Compared with synthetic counterparts, natural pigments have no toxic or side effects on the human body, and even act as nutritional enhancers with various biological activities. As a significant food ingredient, microalgal pigments are also chased by manufacturers and customers. The market value of microalgal pigments has been predicted to be USD 452.4 million by 2025 with a 4% Compound Annual Growth Rate (CAGR), and the market of microalgae products will be a large-scale business until that time (Patel et al., 2022; Sun et al., 2023).

Three major classes of pigments are found in microalgae, carotenoids (usually 0.1–0.2% of dry weight, DW or as high as 14% within certain species), chlorophylls (0.5–1.0% of DW) and phycobiliproteins (PBPs) (8% of DW) (Pereira et al., 2021; Sun et al., 2023).

## 2.1. Chlorophylls

Chlorophylls (with the general formula of $C_{(35-55)}H_{(28-72)}MgN_4O_{(5-7)}$), are greenish pigments discovered in oxygenic photosynthetic organisms including plants, microalgae and cyanobacteria. They are responsible for photosynthesis by absorbing and converting solar energy into chemical energy. At present, five categories of chlorophylls have been identified, including a-f, and their maximal absorption wavelengths ($\lambda_{Amax}$) are, 665, 652, 630, 696 and 707 nm, respectively.

As shown in Figure 1, these categories are composed of four -CH bridged pyrrole rings with a central metal ion (generally a magnesium), enclosed by four nitrogen atoms and connected to the tetrapyrrole ring. This structure contains several associated double bonds, which are responsible for absorbing visible wavelengths, especially at 670-680 nanometers (red) and 435-455 nanometers (blue) (Keykha Akhar et al., 2021; Sun et al., 2023).

Most pigments can bind to several phyla, while chlorophyll a is the most abundant in microalgae (Duppeti et al., 2017; Sun et al., 2023). Chlorophyll b represents the pigment with the second highest abundance within green microalgae; on the other hand, chlorophyll c has been observed within haptophytes, dinoflagellates, cryptophytes and heterokonts. This chlorophyll is divided into three subclasses, which are c1, c2, and c3. C1 is considered to be the most common form of chlorophyll c. Chlorophyll d was identified in Rhodophyta, chlorophyll e (the rarest chlorophyll) is found in *Vaucheria hamata* and *Tribonema bombycinum*. While chlorophyll f is recently discovered in cyanobacteria (Keykha Akhar et al., 2021; Sun et al., 2023).

On average, microalgae contain 0.5-1.0% of chlorophylls per DW. Many microalgae can be the source of natural chlorophylls used for commercial applications, such as *Chlorella* sp., *Monoraphidium dybowskii*, *Scenedesmus dimorphus*, *Chlamydomonas reinhardtii*, *Pavlova lutheri*, *Ankistrodesmus falcatus*, and *Chlorella vulgaris* (Begum et al., 2016; Sun et al., 2023).

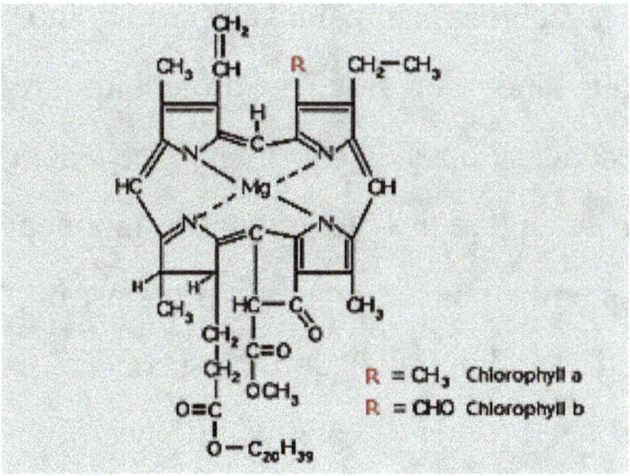

**Figure 1** Chlorophyll a and b (Keykha Akhar et al., 2021).

## 2.2. Carotenoids

Carotenoids are isoprenoids that are synthesized by all photosynthetic organisms and are produced by eight isoprene units ($C_5H_8$) (Figure 2). They have a single long hydrocarbon chain and contain 40 carbon atoms. Carotenoids are water-insoluble pigments and consorted with chloroplast lipids (Mulders et al., 2014; Keykha Akhar et al., 2021).

The different architectures provide carotenoids with outstanding light-absorption characteristics necessary in the photosynthetic process. On the basis of chemical structure, these hydrocarbons are classified into carotenes and xanthophylls. Carotenes, namely hydrocarbons, consist of α- and β-carotene. Xanthophylls contain oxygen as hydroxyl groups (such as zeaxanthin, lutein), keto-groups (such as echinenone, cantaxanthin), or their combination for violaxanthin and astaxanthin. Carotenoids are divided into primary or secondary ones. The primary one's act as the cellular photo- synthetic apparatus' functional components, which are linked with membranes or specific proteins in the thylakoid membrane (e.g., xanthophylls), and the secondary carotenoids are typically produced when exposed to specific environmental stimuli and exist in lipid vesicles (Sun et al., 2023).

Carotenoids have two important effects on photosynthesis: (1) light absorption within visible spectral regions where no chlorophyll is effectively absorbed; (2) photo-protection on photosynthetic systems. Typically, the photoprotective mechanisms have been proven to weaken the energetic state change in chlorophyll induced by excess light radiation absorption. It can effectively hinder reactive oxygen species (ROS) production and endows carotenoids with superior antioxidant capacity (Varela et al., 2015; Sun et al., 2023). Over 600 carotenoids have been identified, while those in microalgae are mainly lycopene, β-carotene, zeaxanthin, astaxanthin, lutein and violaxanthin, which nearly compose 90% of carotenoids within the human body and the diet. Among them, β-carotene, astaxanthin, fucoxanthin and lutein are the most studied ones.

β-carotene is the vitamin A precursor (retinol), which represents an orange-yellow color depending on the quantity of β-carotene and other pigments. Demand for natural β-carotene is increasing, for this carotenoid presents more superior bioactive properties than artificial ones. The microalgae *Dunaliella* contains β-carotene above 14% DW and has long been recognized as the largest natural source. Other microalgae of Cyanobacteria, Chlorophyta, Bacillariophyta, and Euglenophyta also master the capability to produce β-carotene (Sun et al., 2023).

Astaxanthin is a byproduct of β-carotene with a rosy color, it is converted from β-carotene under the action of β-carotene hydroxylase (CRTR-B) and β-carotene oxygenase (CRTO) under stress, followed by accumulation within lipid vesicles out of chloroplasts. In nature, astaxanthin usually exists with 1 or 2 esterified fatty acids (FAs, referred to as monoester and diester, respectively). Astaxanthin can form as different isomers according to hydroxyl group configuration. Microalgae including *Haematococcus pluvialis*, *Chlorococcum* sp. and *Chlorella zofingiensis* have a capability to accumulate astaxanthin in vesicles when subjected to stress (Sun et al., 2020; Sun et al., 2023).

Lutein is a lipophilic tetraterpene, whose color ranges from yellow (low content) to red-orange (high content). The double bonds in lutein endow its high antioxidant activity to be chemically reactive with ROS. On the other hand, lutein can protect photoreceptors by filtering blue light (500 nm). One of the key commercial sources of lutein production is *Muriellopsis* sp., which contains lutein at 0.4–0.6% per dry biomass (Blanco et al., 2007; Sun et al., 2023).

The orange pigment fucoxanthin accounts for more than 10% of total natural carotenoids and is ubiquitous among living organisms in a marine ecosystem. Fucoxanthin has a multi-olefin skeleton and specific conjugated double bonds, as well as single epoxy, carbonyl, and hydroxyl groups (Li et al., 2019; Sun et al., 2023). These unique bioactive structures in fucoxanthin are responsible for its potent health benefits to humans, such as anti-obesity, anti-cancer, and blood glucose regulatory functions. Diatoms (Bacillariophyta, especially *Odontella* sp.) and brown microalgae (Phaeophyceae) have abundant levels of lutein, but their application in industrialized production is still not feasible. Some other microalgae, such as *Phaeodactylum tricornutum*, *Cylindrotheca* sp., *Odontella aurita*, *Chaetoceros muelleri*, *Amphora* sp., *Navicula* sp., and *Chrysotila carterae*, also showed considerable fucoxanthin production capacity (Ishika et al., 2017; Wang et al., 2018; Sun et al., 2023).

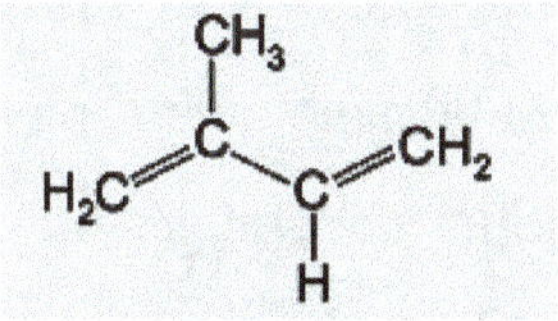

**Figure 2** Isoprene unit (Keykha Akhar et al., 2021).

### 2.3. Phycobiliproteins (PBPs)

PBPs are the main light-harvesting chromoproteins, which capture and pass the light energy on to chlorophylls during photosynthesis and are mainly found in Rhodophyta, Cyanobacteria, and certain type of marine algae. PBPs are water-soluble and highly fluorescent proteins belonging to open-chain tetrapyrroles with linear prosthetic groups (bilins), which are covalently bound via cysteine amino acid residues (Figure 3) (Han et al., 2015; Keykha Akhar et al., 2021; Muñoz-Miranda and Iñiguez-Moreno, 2023). They can be assembled into phycobilisomes or adhered onto thylakoid membrane for photosynthesis (Figure 4) (Keykha Akhar et al., 2021; Chini

Zittelli et al., 2023; Sun et al., 2023). These protein pigments are categorized based on properties such as structure, absorption spectra (consensus maximum absorbance, Lambda max [λmax]), and color.

Based on absorbance wavelength, cyanobacteria and red algae PBPs are classified into four categories, as follows: phycoerythrin (PE) is purple, and its λmax is within the range of 490-570 nanometers; phycoerythrocyanin (PEC) is orange, and its λmax is within the range of 560-600 nanometers; phycocyanin (PC) is blue, and its λmax is within the range of 610-625 nanometers; allophycocyanin (APC) has a bluish green color, and its λmax is within the range of 650-660 nanometers (Keykha Akhar et al., 2021; Sun et al., 2023).

PBPs are also classified into two large groups based on their color; red PBP is known as phycoerythrin, and phycocyanin is blue PBP. Moreover, phycocyanins are divided into three subcategories, which are allophycocyanin (APC), C-phycocyanin (C-PC), and R-phycocyanin (R-PC). PEs are also subdivided into B-PE and R-PE (Manirafasha et al., 2016; Keykha Akhar et al., 2021).

In most cyanobacteria, cryptophytes, glaucophytes and red algae, the common PBP is PC, nowadays, PC can be generated from autotrophic cyanobacteria *Spirulina platensis* because it is ubiquitous. A strain of *Geitlerinema* can also achieve a yield of

172 mg g$^{-1}$ (Ruiz-Domínguez et al., 2021; Sun et al., 2023). While PE can be produced from *Porphyridium* sp. (such as *P. purpureum* and *P. cruentum*), and its content is 15% DW, while its yield is 200 mg L$^{-1}$ (Dufossé et al., 2005; Sun et al., 2023). In comparison to PC and PE, the massive commercial APC manufacturing process has not been established so far because it is limited by the poor intracellular pigment content and biomass generation.

The common biosynthetic pathways of chlorophylls, phycobiliproteins and carotenoids were shown in Figure 5 (Sun et al., 2023).

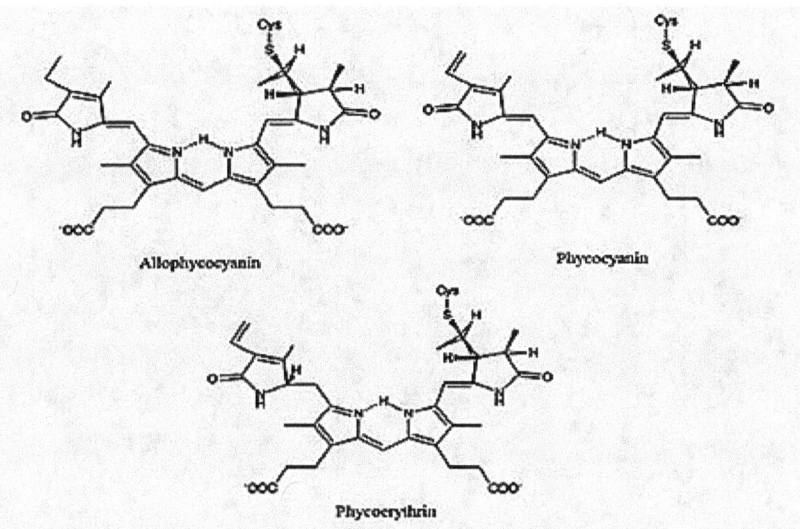

**Figure 3** The chemical structure of phycobiliproteins (Muñoz-Miranda and Iñiguez-Moreno, 2023).

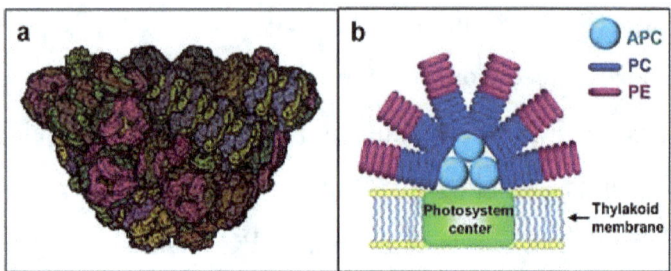

**Figure 4** The Phycobilisome: (a) cryo electron microscopy image (*P. purpureum*) and (b) general structure scheme: APC allophycocyanin, PC phycocyanin and PE phycoerythrin (Chini Zittelli et al., 2023).

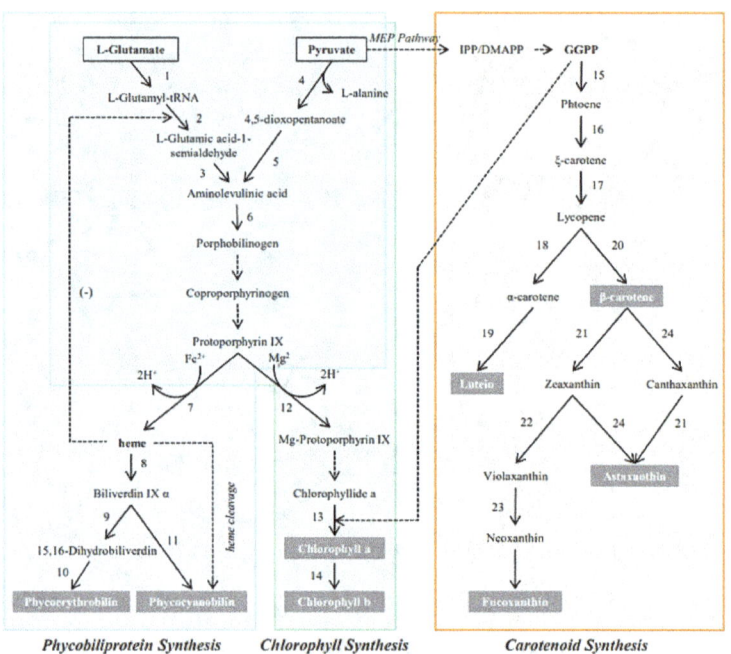

**Figure 5** Biosynthetic pathways of pigments in microalgae. 1: glutamate-tRNA ligase, 2: glutamate-tRNA reductase, 3: 5-aminolevulinate synthase, 4: 5-aminolevulinas, 5: aminolevulinic

acid transaminase, 6: porphobilinogen synthase, 7: ferrochelatase, 8: heme oxgenase, 9: ferrodoxin oxidoreductase (Peb A), 10: ferrodoxin oxidoreductase (Peb B), 11: phycocyanobilin synthase, 12: Mg-protoporphyrinogen IX chelatase, 13: chlorophyll synthetase, 14: chlorophyll a oxygenase (chlorophyll b synthase), 15: phytoene synthase, 16: phytoene desaturase, 17: zeta-carotene isomerase, zeta-carotene desaturase, carotenoid isomerase, 18: lycopene epsilon cyclase, 19: cytochrome P450-β hydroxylase, 20: lycopene-β cyclase, 21: β-carotene hydroxylase, 22: zeaxanthin epoxidase, 23: neoxanthin synthase and 24: β-carotene ketolase. Abbreviations: (IPP) isopentenyl pyrophosphate; (DMAPP) dimethylallyl pyrophosphate; (GGPP) geranylgeranyl pyrophosphate (Sun et al., 2023).

## 3. Factors affecting the microalgal pigment production

Nowadays, the fermentation of microalgae has become one of the effective strategies of natural pigment production. Compared with those from plants and aquatic animals, pigments obtained from the industrial production of microalgae have a multitude of advantages, including controllable production, easy extraction, high yields, no raw materials scarcity, and no seasonal variations. In microalgae, pigments are being produced during vegetative growth as it helps in light-harvesting however, some reports also address their production in stress conditions. Different biotic and abiotic factors such as light irradiance, length of photoperiod, nutrient availability, pH, temperature, salinity, heavy metals, and pesticides, can affect the production of algal pigments. Any slight change in these parameters can cause an alteration of the pigment productivity and molecular structure, ultimately affecting the market acceptance, and the bioaccessibility of final products (Patel et al., 2022; Sun et al., 2023).

## 3.1. Light

Light is the most affecting factor of converting inorganic carbon into organic molecules in phototrophic organisms. Light intensity represents the most obvious and easily regulated factor that affects cell photosynthesis and pigment biosynthesis. Light has a negative effect on photosynthesis under extreme light conditions out of its tolerance limit through the destruction of photosynthetic apparatus. Phycobiliprotein and chlorophyll production is the light collection-related adaptive response. Low light intensity leads to the decreased specific maintenance energy ratio in cyanobacteria, which also stimulates its phycobiliprotein production. Experiments showed that 25-50 μmol photons/m$^2$/s was the highest phycobiliprotein generation intensity in some blue-green microalgae, while that of *Spirulina* increased with the increases in light intensity (135 μmol photons/m$^2$/s) (Madhyastha and Vatsala, 2007; Sun et al., 2023). The chlorophyll content in several microalgae (including *Chlorella* sp., *Dunaliella salina*, *C. reinhardtii*, and *S. platensis*) displays an inverse relationship to light intensity (Chauhan and Pathak, 2010; Sun et al., 2023). In contrast, elevating light intensity shows a positive effect on β-carotene synthesis and accumulation of microalgal cells.

The significance of the light quality on photosynthetic pigments goes beyond that of the light intensity under certain circumstances, and even influences cell maturity, culture density, light path and the medium nutrients profile. Discontinuous illuminating strategies such as the light/dark photoperiod cycle and the flashing light effect have been used to achieve higher light availability. The photoperiod effectively regulates microalgal chlorophyll levels (George et al., 2014; Sun et al., 2023), and usage of the flashing light effect in industrial culture can enhance the astaxanthin generation rate 4-fold within *H. pluvialis* per photon relative to continuous light sources (Kim et al., 2006; Sun et al., 2023).

Light color is also related to pigment generation within diverse microalgae. In some studies, red light is found to favorably enhance PBPs generation within most blue-green microalgae, whereas blue light stimulated that of *Spirulina* sp. and enhanced its chlorophyll production. Green light increases the C-phycoerythrin (C-PE) level in nitrogen-fixation cyanobacteria, and red-light irradiation increases the C-PC level (Hemlata and Fatma, 2009; Sun et al., 2023). The intensity of white light is helpful for synthesizing chlorophyll as well as accumulating C-PC in

*Spirulina* sp., whereas green light positively affects its PC generation (Madhyastha and Vatsala, 2007; Sun et al., 2023).

Light spectrum is the photomorphogenic signal in microalgae cultures. Various lengths of spectrum, including blue: red, red: far red, green: red, and blue: green have been often used at different proportions to achieve the best induction effect. Alteration of spectrum proportions influence the relative pigment composition. The lutein production from *Scenedesmus obliquus* could obtain the maximal productivity (1.43 mg $L^{-1}$ $day^{-1}$) with 4-day blue followed by 4-day red exposure (Chen et al., 2019; Sun et al., 2023). Chlorophyll a content of *Chlorella pyrenoidosa*, declines after blue light irradiation, while it greatly elevates after blue: red irradiation (Wang et al., 2020; Sun et al., 2023). *G. membranacea* and *C. vulgaris* attain highest chlorophyll A levels after red: green light irradiation (Mohsenpour et al., 2012; Sun et al., 2023).

### 3.2. Temperature

Temperature is the elementary factor that can govern the rate of all metabolic processes and cellular component structure by influencing membrane fluidity, enzymatic activities, as well as electron transport chain efficiency. With regard to microalgae, optimal growth and a tolerable temperature range vary among diverse strains (Keykha Akhar et al., 2021; Sun et al., 2023). Typically, high temperatures can promote microalgal pigment production. A range of 25-28 °C was reported optimum in terms of chlorophylls accumulation temperatures, for higher temperatures may stimulate damaging of cells by osmotic pressure (Chauhan and Pathak, 2010; Sun et al., 2023). In blue-green microalgae, the temperature of 25-30°C causes the accumulation of carotenoids (e.g., β-carotene), while the temperature of 28°C facilitates astaxanthin and lutein biosynthesis. The temperature of 25-36°C is considered to be optimal for phycobiliprotein production in various algal strains (Domınguez-Bocanegra et al., 2004; García-González et al., 2005; Keykha Akhar et al., 2021).

### 3.3. Culture media

Several factors may influence microalgal growth rate, biomass, and pigment production. Culture medium and its composition play a pivotal role in this regard.

### 3.3.1. Nitrogen

Nitrogen is an essential macronutrient to microalgal growth and is involved in the synthesis of chlorophyll molecule, carbohydrates, lipids, proteins and nucleic acid. Moreover, nitrogen deficiency has been identified for inducing various cell responses within microalgae, such as stimulating excessive free radical formation. Since carotenoids production has been recognized to be the protective response to photo-oxidative stress resulting from the excessively reduced photosynthetic electron transport chain, nitrogen deficiency can cause a marked increase of their contents. On the other hand, four nitrogen atoms are required to compose four pyrrole groups for the synthesis of each chlorophyll molecule and nitrogen stress displays an adverse effect on its synthesis (Keykha Akhar et al., 2021; Sun et al., 2023).

Nitrate, nitrite, urea, and ammonium are the nitrogen forms that could be used by microalgae. However, using nitrate is more common compared to ammonium salts as it is more stable with the shift of pH, and ammonia concentrations of more than 25 μM are toxic for microalgae (Zhao et al., 2017; Keykha Akhar et al., 2021). It has been reported that *C. vulgaris*, S*cenedesmus subspicatus*, *Chlorella fusca*, *S. platensis* and *C. reinhardtii* have a reduced chlorophyll level and growth dose-dependently as a consequence of nitrogen deficiency (Lv et al., 2010; Sun et al., 2023). On the other hand, Cyanobacteria shows particular requirements for nitrogen sources, many blue-green microalgae (such as *Anabaena* sp.) were proven to produce a high amount of PBPs with nitrogen-free condition, while *Fischerella* sp. showed the opposite tendency (Hemlata and Fatma, 2009; Sun et al., 2023).

### 3.3.2. pH and salinity

Little research is conducted to investigate how pH affects pigment production in microalga, but its alteration can effectively influence nutrient bioavailability and solubility in a culture system (Keykha Akhar et al.,2021; Patel et al., 2022; Sun et al., 2023). In related to the microalgae pigments, the pH 8-9 lead to maximum production of chlorophyll and PBPs. For instance, it has been reported that the pH of 8 led to the optimum production of PBPs in *Anabaena* sp. and *Synechocystis* sp., as well as the production of carotenoids in *Scenedesmus almeriensis*. Furthermore, the pH of 9 is considered optimal for the production of PC, PE, and APC in *Nostoc* sp., as well as the production of chlorophyll in *Spirulina platensis*. The pH value of 6.5- 7.5 is also regarded as the optimum range for carotenoid production; such examples are astaxanthin in *Chlorella zofingiensis*, carotenoid in *Chlorella prototothecoides*, lutein in *Muriellopsis* sp., and *Chlorococcum citriforme*, *Neospongiococcus gelatinosum*, and β-carotene in *Dunaliella salina* (Begum et al., 2016; Keykha Akhar et al.,2021).

Salt concentration and osmotic pressure have a crucial impact on pigment production of both marine and limnetic microalgae. The rapid entry of sodium ions into cells results in phycobilisome detachment from the thylakoid membrane, leading to a significant reduction of photosynthesis and a restriction of pigment production. Thus, excessive salinity is usually not required, for it produces hypertonic solution in the culture liquid and causes cell shrinkage (Keykha Akhar et al., 2021; Sun et al., 2023).

As a general principle, maximum carotenoid production is obtained at high salts concentration, whereas the low concentration of salts increases chlorophyll and PBP production. Previous findings have also indicated that increased salinity could decrease the chlorophyll content, while enhancing the β-carotene production in blue-green microalgae. For instance, *Dunaliella salina* produced 54.12 mg/g of β-carotene at three ppt (parts per trillion) of salinity, and *Oscillatoria* sp. produced 66.70 mg/g of phycoerythrin at 15 ppt. According to Hemlata and Fatma, 2009, PBP production in *Anabaena* showed an increase of up to 135.73 mg/g at 10 ppt of salt (Keykha Akhar et al., 2021; Sun et al., 2023).

### 3.3.3. Micronutrients

Micronutrients have critical effects for pigment metabolic pathways and generally low or even trace amounts are needed (Keykha Akhar et al.,2021; Sun et al., 2023).

Copper serves as an essential cofactor for metalloenzymes in several metabolic pathways but is toxic for microalgae growth at high concentrations. A 1 mg/L copper has been proved to hinder the growth of *I. galbana*, *Pavlova viridis*, and *P. tricornutum*, meanwhile reduce their chlorophyll content. High zinc and copper contents are associated with chloroplast membrane peroxidation caused by free radical production, thus leading to reduction of chlorophyll content. Moreover, magnesium ions, participate in pigment biosynthesis as the cofactor for critical enzymes (Esakkimuthu et al., 2016; Sun et al., 2023). Besides, it has been reported that, iron content of the culture medium positively affects β-carotene and astaxanthin production (Gord-Noshahri et al., 2018; Keykha Akhar et al.,2021).

Furthermore, microalgae require sulfur for producing numerous essential metabolites, such as sulfur-containing amino acids cysteine and methionine. It has been found that sulfur deprivation reduces oxygenic photosynthesis, promotes the activation of hydrogenase and has a negative effect on microalgal chlorophyll accumulation in *C. reinhardtii* and *C. fusca* (Cakmak et al., 2012; Jerez et al., 2016; Sun et al., 2023). On the contrary, sulfur deprivation displays a promoting effect for carotenoids biosynthesis. It is a more efficient approach to induce astaxanthin and lipids accumulation in *H. pluvialis* compared with nitrogen restriction, which was also verified in *C. reinhardtii* and *Parachlorella kessleri* (Cakmak et al., 2012; Yamazaki et al., 2018; Sun et al., 2023).

## 4. Potential applications of microalgal pigments

Microalgal pigments have various applications in biotechnological studies, as well as in food, cosmetics, aquaculture, textiles, nutraceuticals, and pharmaceutical industries. Owing to their nontoxic, non-carcinogenic, antioxidative, and immune-boosting properties, their industrial applications are on the rise. Some of their applications and advantages have been described in the following section.

Chlorophylls have long been used as a traditional therapeutic medicine for wound healing, controlling calcium oxalate crystal, and preventing bacteria advancement. It is acknowledged that these compounds are effective in the treatment of cancer and cardiovascular diseases. Since the chemical structure of chlorophylls is similar to hemoglobin, it may facilitate the rate of carbon dioxide and oxygen interchange. This may explain the positive effects of these compounds on rapid wound healing and the formation of new tissues (Manivasagan et al., 2018; Keykha Akhar et al.,2021). Furthermore, this property renders chlorophyll a key compound in the treatment of postoperative ulcers in rectal surgeries, increasing the rate of healing by 25% (Hosikian et al., 2010; Keykha Akhar et al.,2021). Moreover, chlorophyll derivatives (especially pheophorbide b and pheophytin b) are considered to be potent protective antioxidants, which could scavenge the mutagens in the gastrointestinal tract (Keykha Akhar et al.,2021).

Carotenoids have widespread applications as feed, food/cosmetic additives, and colorants. The positive health properties of carotenoids are mainly associated with their antioxidant activities, which reduce the risk of AIDS, diabetes, cataract, macular degeneration, and neurodegeneration (Varela et al., 2015; Chuyen and Eun, 2017; Keykha Akhar et al.,2021).

Astaxanthin is a significant free radical scavenger, has substantial antioxidant activity, and inhibits lipid peroxidation. The Food and Drug Administration (FDA) has approved the use of astaxanthin as a food colorant in fish and animal feed.

Fucoxanthin is a xanthophyll, which is found in diatoms and golden-brown unicellular microalgae and has anticancer, antioxidant, antiangiogenic, antidiabetic, anti-obesity, anti-inflammatory, and antimalarial properties (Pashkow et al., 2008; Keykha Akhar et al.,2021). On the other hand, previous findings have confirmed the anti-obesity activity of fucoxanthin, especially in combination with pomegranate seed oil (xanthigen). Since obesity is an important public health concern and leads to cardiovascular diseases, hypertension, hyperlipidemia, and type II diabetes, the development of fucoxanthin production has gained importance in an unprecedented manner (Maeda et al., 2009; Hasani-Ranjbar et al., 2013; Keykha Akhar et al.,2021).

Canthaxanthin is a lipid-soluble natural xanthophyll with remarkable antioxidant activity compared to β-carotene (Tanaka et al., 2012; Keykha Akhar et al.,2021). It is used as a food additive for farmed shrimp and salmon fish to improve the skin color. Also, zeaxanthin and lutein are categorized as xanthophylls and have great positive potential in maintaining eye health (Gammone et al., 2015; Keykha Akhar et al.,2021).

Phycocyanin is a water-soluble pigment with potential anticancer activity. It stimulates apoptosis through G2/M cell cycle arrest, and its anticancer activity against human pancreatic adenocarcinoma has been confirmed *in vitro* and *in vivo* (Manivasagan et al., 2018; Keykha Akhar et al.,2021).

The applications of microalgal pigments are summarized in Table 1(Sun et al., 2023).

**Table 1** Microalgal sources, and applications of natural pigments (Sun et al., 2023).

| Pigment | Microalgal source | Bioactivities | References |
|---|---|---|---|
| Chlorophylls | Chlorella sp.<br>Monoraphidium dybowskii<br>Scenedesmus dimorphus<br>Chlamydomonas reinhardtii<br>Pavlova lutheri<br>Chlorella vulgaris | Improving immune system, antioxidant, and anti-carcinogen | (Jurić et al., 2022) |
| β-Carotene | Dunaliella salina<br>Dunaliella bardawil | Anticancer, antioxidant, cholesterol synthesis suppressor, skin protectant against sunburn, prevent coronary artery diseases, fatty liver disease, type-2 diabetes, insulin resistance, and age-related macular degeneration, and | (Sluijs et al., 2015; Beydoun et al., 2019) |

| | | | |
|---|---|---|---|
| | | ultraviolet-induced skin cancer and oral carcinomas | |
| **Astaxanthin** | *Haematococcus pluvialis* *Chlorella zofingiensis* *Chlorella sp.* | Antioxidant and photoprotector, anti-inflammatory, antineoplastic, anticancer, antimicrobial, and anti-hyperlipidemia Increase serum adiponectin Beneficial effects on blood rheology and metabolic syndrome | (Chuyen and Eun, 2017; Pogorzelska et al., 2018) |
| **Lutein** | *Chlorella protothecoides* *Muriellopsis sp.* *Scenedesmus almeriensis* *Dunaliella salina* *Dunaliella tertiolecta* *Brassica oleracea* *Spinacia oleracea* | Antioxidant, prevent the production of free radicals, immunity strengthens, anti-inflammatory, anti-atherogenic, and antihypertensive Cytoprotection against alcohol-induced liver injury Potent neuroprotection | (Sun et al., 2014; Shakeri et al., 2018) |

| | | | |
|---|---|---|---|
| Fucoxanthin | *Actinidia deliciosa* | Prevention of cataracts, age-related macular degeneration, skeletal ischemia, hepatotoxicity and cardiovascular diseases | (Park et al., 2010; Jurić et al., 2022) |
| | *Phaeodactylum tricornutum*<br>*Cylindrotheca closterium* | Reduce blood triglyceride concentration, anti-inflammatory, inhibit pro-inflammatory factors, improve phagocytic and microbicidal capacity | |
| | *Isochrysis galbana* | Antioxidant, decrease oxidative damage to lipids/proteins | |
| | *Mallomonas sp.* | Antineoplastic, inhibit the growth of human leukemia cells and neuroblastoma cells | |
| | *Nitzschia laevis*<br>*Odontella aurita*<br>*Chaetoceros sp.* | Anti-obesity | |

| Phycobiliproteins | *Spirulina platensis*<br>*Geitlerinema*<br>*Porphyridium sp.* | Anti-alzhelmeric activity, antioxidant | (Cuellar-Bermudez et al., 2015; Jacob-Lopes et al., 2019) |

## 5. Future perspective and conclusive remarks

With the exalting of civilization, man-kind has realized the significance of products obtained from natural feedstocks. When renewability becomes a crucial factor for the commercial production of food, feed, and fuel, algal biomass holds a prominent standpoint in the global market.

Due to the significantly higher bioactive potential of microalgal pigments compared to their synthetic counterparts, microalgal pigments demand is rising as food additives, feed, nutrient supplements, and dyes. Moreover, it has been impressed recently for application in nutraceutical, cosmeceutical, and pharmaceuticals. Furthermore, these pigments find broader utilization as antioxidant, anticancer, and anti-inflammatory agents. The remarkable credentials of algal pigments upsurged their value in the global market. However, the production of most microalgal pigments is still confined to the experimental stage, and its commercial industrialization is far from reality. Thus, more research is required to dig untapped potential technologies, referring to microalgae characteristics, pigment properties, and biosynthetic metabolism, hence defining further applications.

# References

Begum, H., Yusoff, F.M., Banerjee, S., Khatoon, H. and Shariff, M., 2016. Availability and utilization of pigments from microalgae. Critical Reviews in Food Science and Nutrition. 56, 2209-2222.

Beydoun, M.A., Chen, X., Jha, K., Beydoun, H.A., Zonderman, A.B. and Canas, J.A., 2019. Carotenoids, vitamin A, and their association with the metabolic syndrome: A systematic review and meta-analysis. Nutrition Reviews. 77, 32-45.

Blanco, A.M., Moreno, J., Del Campo, J.A., Rivas, J. and Guerrero, M.G., 2007. Outdoor cultivation of lutein-rich cells of *Muriellopsis* sp. in open ponds. Applied Microbiology and Biotechnology. 73, 1259-1266.

Cakmak, T., Angun, P., Demiray, Y.E., Ozkan, A.D., Elibol, Z. and Tekinay, T., 2012. Differential effects of nitrogen and sulfur deprivation on growth and biodiesel feedstock production of *Chlamydomonas reinhardtii*. Biotechnology and Bioengineering. 109, 1947-1957.

Chauhan, U.K. and Pathak, N., 2010. Effect of different conditions on the production of chlorophyll by *Spirulina platensis*. J Algal Biomass Utln. 1, 89-99.

Chen, W.C., Hsu, Y.C., Chang, J.S., Ho, S.H., Wang, L.F. and Wei, Y.H., 2019. Enhancing production of lutein by a mixotrophic cultivation system using microalga *Scenedesmus obliquus* CWL-1. Bioresource Technology. 291, p.121891.

Chini Zittelli, G., Lauceri, R., Faraloni, C., Silva Benavides, A.M. and Torzillo, G., 2023. Valuable pigments from microalgae: phycobiliproteins, primary carotenoids, and fucoxanthin. Photochemical & Photobiological Sciences. 22, 1733-1789.

Chuyen, H.V. and Eun, J.B., 2017. Marine carotenoids: Bioactivities and potential benefits to human health. Critical Reviews in Food Science and Nutrition. 57, 2600-2610.

Cuellar-Bermudez, S.P., Aguilar-Hernandez, I., Cardenas-Chavez, D.L., Ornelas-Soto, N., Romero-Ogawa, M.A. and Parra-Saldivar, R., 2015. Extraction and purification of high-value metabolites from microalgae: essential lipids, astaxanthin and phycobiliproteins. Microbial Biotechnology. 8, 190-209.

Dominguez-Bocanegra, A.R., Legarreta, I.G., Jeronimo, F.M. and Campocosio, A.T., 2004. Influence of environmental and nutritional factors in the production of astaxanthin from *Haematococcus pluvialis*. Bioresource Technology. 92, 209-214.

Dufossé, L., Galaup, P., Yaron, A., Arad, S.M., Blanc, P., Murthy, K.N.C. and Ravishankar, G.A., 2005. Microorganisms and microalgae as sources of pigments for food use: a scientific oddity or an industrial reality?. Trends in Food Science & Technology. 16, 389-406.

Duppeti, H., Chakraborty, S., Das, B.S., Mallick, N. and Kotamreddy, J.N.R., 2017. Rapid assessment of algal biomass and pigment contents using diffuse reflectance spectroscopy and chemometrics. Algal Research. 27, 274-285.

Esakkimuthu, S., Krishnamurthy, V., Govindarajan, R. and Swaminathan, K., 2016. Augmentation and starvation of calcium, magnesium, phosphate on lipid production of *Scenedesmus obliquus*. Biomass and Bioenergy. 88, 126-134.

Fu, W., Nelson, D.R., Mystikou, A., Daakour, S. and Salehi-Ashtiani, K., 2019. Advances in microalgal research and engineering development. Current Opinion in Biotechnology. 59, 157-164.

Gammone, M.A., Riccioni, G. and D'Orazio, N., 2015. Marine carotenoids against oxidative stress: effects on human health. Marine Drugs. 13, 6226-6246.

García-González, M., Moreno, J., Manzano, J.C., Florencio, F.J. and Guerrero, M.G., 2005. Production of *Dunaliella salina* biomass rich in 9-cis-β-carotene and lutein in a closed tubular photobioreactor. Journal of Biotechnology. 115, 81-90.

George, B., Pancha, I., Desai, C., Chokshi, K., Paliwal, C., Ghosh, T. and Mishra, S., 2014. Effects of different media composition, light intensity and photoperiod on morphology and physiology of freshwater microalgae *Ankistrodesmus falcatus*- A potential strain for bio-fuel production. Bioresource Technology. 171, 367-374.

Gord-Noshahri, N., Ameri, M. and Jalal Ghasem, B., 2018. *Spirulina* production in different sources of nitrogen. Journal of Phycological Research. 2, 145-153.

Han, C., Takayama, S. and Park, J., 2015. Formation and manipulation of cell spheroids using a density adjusted PEG/DEX aqueous two- phase system. Scientific Reports. 5, 1-12.

Hasani-Ranjbar, S., Jouyandeh, Z. and Abdollahi, M., 2013. A systematic review of anti-obesity medicinal plants-an update. Journal of Diabetes & Metabolic Disorders. 12, 1-10.

Hemlata and Fatma, T., 2009. Screening of cyanobacteria for phycobiliproteins and effect of different environmental stress on its yield. Bulletin of Environmental Contamination and Toxicology. 83, 509-515.

Hosikian, A., Lim, S., Halim, R. and Danquah, M.K., 2010. Chlorophyll extraction from microalgae: A review on the process engineering aspects. International Journal of Chemical Engineering. 2010, 1-11.

Ishika, T., Moheimani, N.R., Bahri, P.A., Laird, D.W., Blair, S. and Parlevliet, D., 2017. Halo-adapted microalgae for fucoxanthin production: Effect of incremental increase in salinity. Algal Research. 28, 66-73.

Jacob-Lopes, E., Maroneze, M.M., Deprá, M.C., Sartori, R.B., Dias, R.R. and Zepka, L.Q., 2019. Bioactive food compounds from microalgae: An innovative framework on industrial biorefineries. Current Opinion in Food Science. 25, 1-7.

Jcrcz, C.G., Malapascua, J.R., Sergejevová, M., Figueroa, F.L. and Masojídek, J., 2016. Effect of nutrient starvation under high irradiance on lipid and starch accumulation in *Chlorella fusca* (Chlorophyta). Marine Biotechnology. 18, 24-36.

Jurić, S., Jurić, M., Król-Kilińska, Ż., Vlahoviček-Kahlina, K., Vinceković, M., Dragović-Uzelac, V. and Donsì, F., 2022. Sources, stability, encapsulation and application of natural pigments in foods. Food Reviews International. 38, 1735-1790.

Keykha Akhar, F., Fakhrfeshani, M., Alipour, H. and Ameri, M., 2021. Microalgal pigments: An introduction to their biosynthesis, applications and genetic engineering. Journal of Plant Molecular Breeding. 9, 42-61.

Kim, Z.H., Kim, S.H., Lee, H.S. and Lee, C.G., 2006. Enhanced production of astaxanthin by flashing light using *Haematococcus pluvialis*. Enzyme and Microbial Technology. 39, 414-419.

Li, Y., Sun, H., Wu, T., Fu, Y., He, Y., Mao, X. and Chen, F., 2019. Storage carbon metabolism of *Isochrysis zhangjiangensis* under different light intensities and its application for co-production of fucoxanthin and stearidonic acid. Bioresource Technology. 282, 94-102.

Lv, J.M., Cheng, L.H., Xu, X.H., Zhang, L. and Chen, H.L., 2010. Enhanced lipid production of *Chlorella vulgaris* by adjustment of cultivation conditions. Bioresource Technology. 101, 6797-6804.

Madhyastha, H.K. and Vatsala, T.M., 2007. Pigment production in *Spirulina fussiformis* in different photophysical conditions. Biomolecular Engineering. 24, 301-305.

Maeda, H., Hosokawa, M., Sashima, T., Murakami-Funayama, K. and Miyashita, K., 2009. Anti-obesity and anti-diabetic effects of fucoxanthin on diet-induced obesity conditions in a murine model. Molecular Medicine Reports. 2, 897-902.

Manirafasha, E., Ndikubwimana, T., Zeng, X., Lu, Y. and Jing, K., 2016. Phycobiliprotein: Potential microalgae derived pharmaceutical and biological reagent. Biochemical Engineering Journal. 109, 282-296.

Manivasagan, P., Bharathiraja, S., Santha Moorthy, M., Mondal, S., Seo, H., Dae Lee, K. and Oh, J., 2018. Marine natural pigments as potential sources for therapeutic applications. Critical Reviews in Biotechnology. 38, 745-761.

Mohsenpour, S.F., Richards, B. and Willoughby, N., 2012. Spectral conversion of light for enhanced microalgae growth rates and photosynthetic pigment production. Bioresource Technology. 125, 75-81.

Mulders, K.J., Lamers, P.P., Martens, D.E. and Wijffels, R.H., 2014. Phototrophic pigment production with microalgae: biological constraints and opportunities. Journal of Phycology. 50, 229-242.

Muñoz-Miranda, L.A. and Iñiguez-Moreno, M., 2023. An extensive review of marine pigments: sources, biotechnological applications, and sustainability. Aquatic Sciences. 85, 1-21.

Park, J.S., Chyun, J.H., Kim, Y.K., Line, L.L. and Chew, B.P., 2010. Astaxanthin decreased oxidative stress and inflammation and enhanced immune response in humans. Nutrition & Metabolism. 7, 1-10.

Pashkow, F.J., Watumull, D.G. and Campbell, C.L., 2008. Astaxanthin: a novel potential treatment for oxidative stress and inflammation in cardiovascular disease. The American Journal of Cardiology. 101, S58-S68.

Patel, A.K., Albarico, F.P.J.B., Perumal, P.K., Vadrale, A.P., Nian, C.T., Chau, H.T.B., Anwar, C., ud din Wani, H.M., Pal, A., Saini, R. and Senthilkumar, B., 2022. Algae as an emerging source of bioactive pigments. Bioresource Technology. 351, 1-15.

Pereira, A.G., Otero, P., Echave, J., Carreira-Casais, A., Chamorro, F., Collazo, N., Jaboui, A., Lourenço-Lopes, C., Simal-Gandara, J. and Prieto, M.A., 2021. Xanthophylls from the sea: algae as source of bioactive carotenoids. Marine Drugs. 19, 1-31.

Pogorzelska, E., Godziszewska, J., Brodowska, M. and Wierzbicka, A., 2018. Antioxidant potential of *Haematococcus pluvialis* extract rich in astaxanthin on colour and oxidative stability of raw ground pork meat during refrigerated storage. Meat Science. 135, 54-61.

Ruiz-Domínguez, M.C., Fuentes, J.L., Mendiola, J.A., Cerezal-Mezquita, P., Morales, J., Vílchez, C. and Ibáñez, E., 2021. Bioprospecting of cyanobacterium in Chilean coastal desert, *Geitlerinema* sp. molecular identification and pressurized liquid extraction of bioactive compounds. Food and Bioproducts Processing. 128, 227-239.

Shakeri, A., Soheili, V., Karimi, M., Hosseininia, S.A. and Fazly Bazzaz, B.S., 2018. Biological activities of three natural plant pigments and their health benefits. Journal of Food Measurement and Characterization. 12, 356-361.

Show, P.L., Tang, M.S., Nagarajan, D., Ling, T.C., Ooi, C.W. and Chang, J.S., 2017. A holistic approach to managing microalgae for biofuel applications. International Journal of Molecular Sciences. 18, 1-34.

Sluijs, I., Cadier, E., Beulens, J.W.J., Spijkerman, A.M.W. and Van der Schouw, Y.T., 2015. Dietary intake of carotenoids and risk of type 2 diabetes. Nutrition, Metabolism and Cardiovascular Diseases. 25, 376-381.

Sreenikethanam, A., Raj, S., Gugulothu, P. and Bajhaiya, A.K., 2022. Genetic engineering of microalgae for secondary metabolite production: Recent developments, challenges, and future prospects. Frontiers in Bioengineering and Biotechnology. 10, 1-14.

Sun, H., Li, X., Ren, Y., Zhang, H., Mao, X., Lao, Y., Wang, X. and Chen, F., 2020. Boost carbon availability and value in algal cell for economic deployment of biomass. Bioresource Technology. 300, p.122640.

Sun, H., Wang, Y., He, Y., Liu, B., Mou, H., Chen, F. and Yang, S., 2023. Microalgae-derived pigments for the food industry. Marine Drugs. 21,1-27.

Sun, Y.X., Liu, T., Dai, X.L., Zheng, Q.S., Hui, B.D. and Jiang, Z.F., 2014. Treatment with lutein provides neuroprotection in mice subjected to transient cerebral ischemia. Journal of Asian Natural Products Research. 16, 1084-1093.

Tanaka, T., Shnimizu, M. and Moriwaki, H., 2012. Cancer chemoprevention by caroteno. Molecules. 17, 3202-3242.

Uma, V.S., Usmani, Z., Sharma, M., Diwan, D., Sharma, M., Guo, M., Tuohy, M.G., Makatsoris, C., Zhao, X., Thakur, V.K. and Gupta, V.K., 2023. Valorisation of algal biomass to value-added metabolites: Emerging trends and opportunities. Phytochemistry Reviews. 22,1015-1040.

Varela, J.C., Pereira, H., Vila, M. and León, R., 2015. Production of carotenoids by microalgae: achievements and challenges. Photosynthesis Research. 125, 423-436.

Wang, H., Zhang, Y., Chen, L., Cheng, W. and Liu, T., 2018. Combined production of fucoxanthin and EPA from two diatom strains *Phaeodactylum tricornutum* and *Cylindrotheca fusiformis* cultures. Bioprocess and Biosystems Engineering. 41, 1061-1071.

Wang, X., Zhang, P., Wu, Y. and Zhang, L., 2020. Effect of light quality on growth, ultrastructure, pigments, and membrane lipids of *Pyropia haitanensis*. Journal of Applied Phycology. 32, 4189-4197.

Yamazaki, T., Konosu, E., Takeshita, T., Hirata, A., Ota, S., Kazama, Y., Abe, T. and Kawano, S., 2018. Independent regulation of the lipid and starch synthesis pathways by sulfate metabolites in the green microalga *Parachlorella kessleri* under sulfur starvation conditions. Algal Research. 36, 37-47.

Zhao, L.S., Li, K., Wang, Q.M., Song, X.Y., Su, H.N., Xie, B.B., Zhang, X.Y., Huang, F., Chen, X.L., Zhou, B.C. and Zhang, Y.Z., 2017. Nitrogen starvation impacts the photosynthetic performance of *Porphyridium cruentum* as revealed by chlorophyll a fluorescence. Scientific Reports. 7, p.8542.

# CHAPTER VI
# Analytical Insights into Natural Marine Organic Matter: Compositional and Structural Characteristics

**Hussein Oraby,** *PhD*

Department of Chemical Engineering, Military Technical College, Cairo, Egypt.

### Abstract

Natural marine Organic Matter (OM) constitutes a multifaceted amalgamation of carbohydrates, lipids, and proteins dispersed throughout seawater and sediment matrices, exerting considerable influence on various processes within marine ecosystems, including the cycling of nutrients and the sustenance of living organisms. Given the intricate nature of environmental inquiries pertaining to OM composition, the imperative for employing precise analytical methodologies arises. Consequently, this chapter aims to expound upon the predominant and contemporary approaches utilized for OM characterization, encompassing the entirety of analytical procedures from initial sample preparation to conclusive instrumental analysis employing spectroscopic and chromatographic methodologies. Furthermore, recent advancements pertaining to the elucidation of OM's structural attributes are elucidated.

**Keywords:** marine organic matter; extraction; purification; spectroscopic methods; chromatographic methods.

# 1. Introduction

The predominant portion of organic carbon residing on the Earth's surface is situated within seawater, manifesting primarily as Dissolved Organic Matter (DOM) **(Wells, 2002)**. Additionally, organic matter exists in an undissolved state along the water column as Particulate Organic Matter (POM) and within sedimentary deposits. Within the marine milieu, organic matter constitutes a complex amalgamation of carbohydrates, lipids, and proteins, exhibiting diverse physicochemical structures, configurations, and dimensions. Notably, organic matter encompasses Fulvic Acids (FAs), soluble components across a broad pH range, Humic Acids (HAs), soluble under alkaline conditions, and humin, insoluble irrespective of pH, following definitions commonly utilized in terrestrial (Schulten and Schnitzer, 1995) and aquatic (Ishiwatari, 1992) environments. The varied physical states of organic matter within the marine ecosystem undergo numerous chemical transformations, creating a continuum encompassing DOM, POM, FAs, HAs, and humin. These transformations intricately influence microbial processes, sedimentation dynamics, biogeochemical cycles, marine carbohydrate chemistry, and particle behavior in oceanic systems (Verdugo, 2004).

The efficacy and credibility of studies pertaining to organic matter hinge upon the judicious selection of analytical methodologies tailored to analyze the diverse chemical constituents within the marine environment. Consequently, the objective of this chapter is to delineate and deliberate upon the preeminent analytical methodologies employed for the chemical and structural characterization of organic matter. This review scrutinizes the analytical techniques utilized for discerning the myriad classes of chemical compounds within organic matter, encompassing both contemporary and traditional methodologies while assessing their respective advantages and limitations.

## 2. Extraction, Separation, and Purification Techniques for Organic Matter Isolation from Seawater and Sediments

Extraction methodologies play a pivotal role in environmental investigations concerning OM analysis. The efficacy of extraction procedures relies on the inherent heterogeneity of marine OM and the ensuing molecular interactions, encompassing both polar and non-polar interactions among various chemical constituents (Schulten and Schnitzer, 1995; Piccolo, 2001; Verdugo, 2004). This underscores the significance of meticulously selecting separation and purification methodologies aligned with the chosen extraction techniques. Most extraction methods exhibit specificity towards particular chemical classes (e.g., proteins, carbohydrates, lipids) targeted for analysis, a topic that will be expounded upon in subsequent sections of this chapter. Within this section, we delineate the extraction methods employed for structural characterization of OM derived from both aquatic environments and sedimentary matrices. Numerous extraction solutions have been devised to extract Dissolved Organic Matter (DOM) from marine sediments, typically leveraging an alkaline medium to capitalize on the abundance of hydrophilic functional groups present in the dissolved OM fraction of Fulvic Acids (FAs) and Humic Acids (HAs). Various agents such as sodium/potassium hydroxide and sodium pyrophosphate have been utilized (Garcia et al., 1993; Belzile et al., 1997), alongside saline solutions and organic solvents to target specific compounds within OM (Senesi et al., 1994; Petronio, 1997; Zitko, 2001). Nevertheless, studies indicate that sodium hydroxide yields superior OM recoveries, leading to its endorsement as an official method by the International Humic Substance Society (IHSS). The IHSS protocol involves sequential 1 M HCl pre-treatments preceding each 1 M NaOH extraction, aimed at dissolving metal fractions associated with hydrophilic groups; each extraction step typically spans 24 hours. However, this method entails laborious and time-consuming procedures, often necessitating more than ten extractions to accommodate the disparate solubility characteristics of OM chemicals (Senesi et al., 1989).

Alternative approaches to the IHSS methodology have been explored to enhance the homogeneity of the extracted qualitative composition. For instance, substitution of 6 M HCl for 1 M HCl has been proposed successfully (Mecozzi et al., 1998). Additionally, modifications incorporating ultrasound have been investigated to expedite extraction procedures; the initial ultrasound-assisted method reduces the duration of NaOH steps from 24 hours to 30 minutes (Mecozzi et al., 2002c). Subsequent enhancements to the ultrasound technique by Moredo-Piñeiro et al. (2004, 2006) further reduce the duration of HCl treatments.

Purification and pre-concentration steps are invariably integrated with extraction procedures to eliminate inorganic salts from OM extracts, necessitating the utilization of suitable resins. Commonly employed resins include XAD2, XAD4, and XAD8, each exhibiting distinct recovery performances. Comparative analyses reveal that XAD8 yields the most favorable recoveries (Esteves et al., 1995), thus establishing its widespread adoption (Vojvodić and Ćosović, 1996; Calace et al., 2006a, 2006b). Conversely, when employing other XAD resins, a substantial portion of hydrophobic organic matter, up to approximately 32%, may remain uneluted unless subjected to rigorous hydrolysis procedures (Lara and Thomas, 1994). Numerous investigations have been conducted to explore purification methodologies utilizing alternative XAD8 resins. A recent study indicates that the purification efficacy of XAD8 closely parallels that of the CG300 m resin in OM purification (Pietrantonio et al., 2003). Furthermore, the combined utilization of XAD2 resin with the ion exchange Chelex 100 resin demonstrates comparable purification efficacy to the conventional pre-acidification step typically employed with XAD8 resin (Slauenwhite and Wargersky, 1996). Alternative methods for OM purification involve the utilization of $C_{18}$ Solid-Phase Extraction (SPE), extensively reviewed for environmental analysis by Liška (2000). Another strategy entails the combined application of $C_{18}$ SPE with ultrafiltration, resulting in enhanced recoveries compared to those achieved with $C_{18}$ SPE alone (Simjouw et al., 2005).

Ultrafiltration offers distinct advantages over purification methods reliant on dialysis. By judiciously selecting cut-off membranes, it becomes feasible to segregate various molecular ranges of OM, facilitating specific investigations into OM abundance, size distribution, and aggregation mechanisms (Guo and Santschi, 1996; Benner et al., 1997; Guo et al., 2000; Mecozzi and Pietrantonio, 2006). These investigations encompass conventional, tangential, and cross-flow ultrafiltration techniques.

## 3. Carbohydrate analysis

### 3.1 Total carbohydrate analysis

Total carbohydrate analysis involves the comprehensive categorization of various sub-classes, including neutral sugars, uronic acids, aminosugars, and deoxysugars, facilitated by spectrophotometric techniques operating within the visible region. Figure 1 demonstrates te chemical structure of carbohydrate. Among these techniques, the Phenol Sulphuric Acid Method (PSAM) stands out as one of the most utilized and traditional approaches, applicable to both liquid and solid samples directly (Dubois et al., 1956). However, contemporary methods such as 3-methyl-2-benzothiazolinone hydrazone hydrochloride (MBTH) (Johnson and Sieburth, 1977) and 2, 4, 6-tripyridyl-s-triazine (TPTZ) (Myklestad et al., 1997) are increasingly employed. Each method presents distinct advantages and limitations relative to others, necessitating consideration of various analytical factors when selecting the appropriate methodology.

The PSAM method offers the simplest handling procedure compared to MBTH and TPTZ methods. In PSAM, all derivatization steps for color development can be carried out in a single step at (25–30) °C, whereas the latter two methods involve multiple steps, rendering them more time-consuming. However, PSAM exhibits lower sensitivity (approximately one order of magnitude) than MBTH and TPTZ methods. Consequently, for seawater analysis, the use of a 10 cm path length cell is strongly advised to achieve a detection limit comparable to that of MBTH (Mecozzi et al., 1999). An alternative method for carbohydrate analysis in marine samples is the tryptophan Sulphuric acid method (Josefsson et al., 1972), albeit rarely utilized due to its inherent low analytical reproducibility and accuracy.

The analytical accuracy of each technique for determining total carbohydrates is significantly influenced by the inherent heterogeneity of the carbohydrate fraction in marine samples. Notably, all aforementioned techniques utilize glucose for instrumental calibration due to its prevalence in marine samples. However, each subclass of sugars exhibits a distinct chemical response (i.e., spectral sensitivity) to the derivatization reagent employed. In the PSAM technique, the signal to concentration ratio of uronic acids ranges from 30% to 40% of that of neutral carbohydrates like glucose, with aminosugars showing no reaction (Grasshoff et al., 1983). Conversely, in the TPTZ technique, uronic acids exhibit a signal to concentration ratio ranging from 75% to 55% of the signal to concentration ratio of neutral carbohydrates (Myklestad et al., 1997). Notably, the literature lacks investigations into differences among the spectral characteristics of neutral, uronic acids, and aminosugars in the MBTH technique. The presence of substantial amounts of uronic acids and aminosugars in marine samples can lead to reduced analytical accuracy due to differences in spectroscopic characteristics between standard and real samples. A strategy to enhance analytical accuracy in heterogeneous samples involves combining one of the aforementioned methods for total carbohydrate analysis with specific spectrophotometric methods for determining total uronic acids (Blumenkrantz and Asboe-Hansen, 1973) or aminosugars (Belcher et al., 1954). These methods exhibit high selectivity as neutral sugars do not react under the experimental conditions employed. This approach has been applied in environmental studies focusing on OM aggregates (Giani et al., 2005; Mecozzi et al., 2005).

Uronic acids can also be determined using a modification of the sulphamate/m-hydroxyldiphenyl assay (Filisetti-Cozzi and Carpita, 1991), incorporating cation-exchange separation and lyophilization steps to enhance analytical sensitivity and accuracy (Hung and Santschi, 2001).

Another approach to improving analytical accuracy in total carbohydrate analysis involves modifying the PSAM technique (Mecozzi, 2005). In this method, standard solutions containing glucose, glucuronic acid, and amino glucose are utilized simultaneously, followed by spectroscopic calibration using multivariate methods such as Principal Component Regression (PCR) or Partial Least Squares Regression (PLSR). These mathematical approaches enable better description of spectral patterns arising from heterogeneous mixtures of carbohydrates present in environmental samples.

The analytical accuracy of total carbohydrate estimation is significantly influenced by the hydrolysis step preceding chemical derivatization for spectroscopic determination, regardless of the method employed. Carbohydrates in organic OM originate from phytoplanktonic organisms and algal plants subjected to various degradation–recondensation reactions (Ishiwatari, 1992; Borch and Kirchman, 1997), leading to their presence as mono, oligo, and polysaccharides, while spectrophotometric determination necessitates their quantitative conversion into monosaccharides. Various hydrolytic treatments have been utilized, with 1 M HCl treatment at 100°C for 20 hours being commonly employed (Johnson and Sieburth, 1977), while others opt for sulfuric acid treatment at 100°C (Pakulski and Benner, 1992). However, recent experimental findings indicate the presence of several products resulting from oxidative reactions of carbohydrates in both HCl and $H_2SO_4$ solutions (Ledl and Schleicher, 1990; Mecozzi et al., 2002a), thus affecting analytical accuracy. Consequently, alternative hydrolytic methods have been explored.

Hydrolysis with 12 M $H_2SO_4$ at room temperature for 24 hours yields more quantitative recoveries compared to high-temperature HCl and $H_2SO_4$ treatments (Borch and Kirchman, 1997), aligning well with some ultrasound-assisted hydrolytic methods recently proposed (Mecozzi et al., 1999, 2002a). The ultrasound method offers the advantage of being less time-consuming, particularly when coupled with the PSAM method. Additionally, ultrasound-assisted hydrolysis can be applied to carbohydrate determination in marine sediments (Mecozzi et al., 2000).

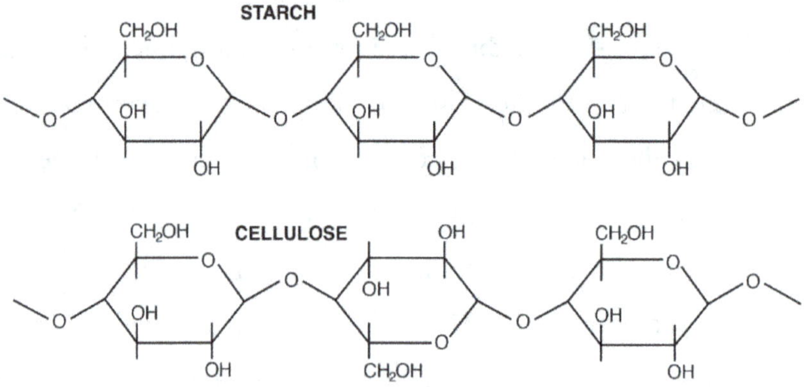

**Figure 1** Chemical structure of carbohydrates

## 3.2 Analysis of monosaccharide composition

The analysis of monosaccharide composition within OM through chromatographic methods serves as a valuable tool for studying cellular components in living organisms and for characterizing the structural composition of polymeric exudates in marine colloids (Verdugo et al., 2004). Regarding gas chromatography (GC) methods, the conversion of sugars into volatile compounds is achieved through the preparation of alditol acetate derivatives (Sawardeker et al., 1965; Oades, 1967; Walters and Hedges, 1988) or by the preparation of trimethylsilyl derivatives (McCarthy et al., 1996). Figure 2 shows the chemical structure of monosaccharide.

When the initial hydrolytic treatment is quantitative, chromatographic methods become reliable, and they can be further enhanced by employing Flame Ionization Detection (FID) and Mass Spectrometry (MS) for detection. However, the derivatization step in these methods can be time-consuming. To address this limitation, alternative procedures that eliminate the need for specific derivatization steps have been developed using liquid and planar chromatography techniques.

In liquid chromatography, neutral sugars (Borch and Kirchman, 1997) and amino sugars (Kaiser and Benner, 2000) are determined through high-performance anion-exchange chromatography with pulsed amperometric detection. Prior to injection, both methods require the removal of inorganic salts from the solutions using strong cation resins.

For planar chromatography, rapid methods for determining monosaccharide composition in OM aggregates have been developed. These include paper chromatography with fluorescence detection (Han and Robyt, 1998), thin-layer chromatography with FID detection (Mecozzi and Pápai, 2004), and high-performance thin-layer chromatography (Marsit et al., 2000; Doner, 2001).

Lastly, specific environmental studies focused on understanding the mechanism of OM aggregation necessitate the determination of polymerized carbohydrates, defined as Transparent Exo-Polymeric Particles (TEP). The quantitative determination of these polysaccharides is conducted in the filtered fraction of seawater samples after staining with Alcian blue dye and detection at 787 nm (Passow and Alldredge, 1995).

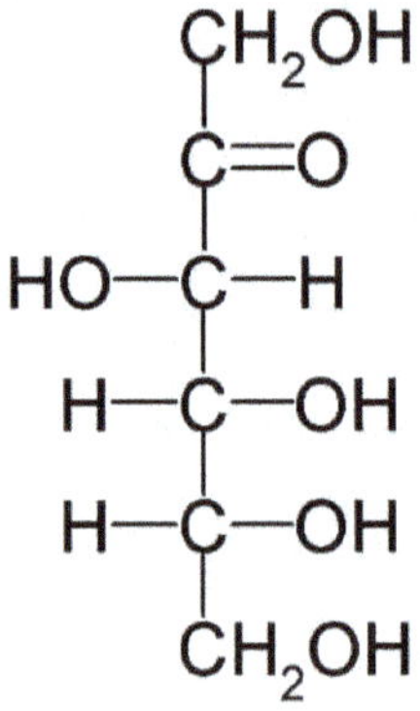

$$\begin{array}{c} CH_2OH \\ | \\ C=O \\ | \\ HO-C-H \\ | \\ H-C-OH \\ | \\ H-C-OH \\ | \\ CH_2OH \end{array}$$

**Figure 2** Chemical structure of monosaccharide

### 4. Lipid analysis

Lipids constitute perhaps the most diverse fraction of OM in marine samples, encompassing hydrocarbons (saturated and unsaturated), fatty acids, ester fatty acids, sterols, alcohols, wax, phospholipids, and chlorophyll pigments (Parrish, 1988). Consequently, the determination of total lipid content typically involves an extraction step followed by gravimetric detection, whereas chromatographic methods are often employed for analyzing specific lipid subclasses (Lin and McKeon, 2005).

Common extraction procedures for total lipids in environmental analysis include the methods developed by Folch et al. (1957) and Bligh and Dyer (1959), renowned for their proven qualitative and quantitative reliability, yielding recoveries higher than those typically obtained by Soxhlet extraction (Ewald et al., 1998).

Alternative methods based on ultrasound-assisted extraction allow for reduced extraction time without compromising analytical accuracy. However, careful selection of experimental conditions such as solvent, time, ultrasound bath temperature, and frequencies is essential to prevent degradative modifications of extracted lipids (Koh, 1983; Mecozzi et al., 2002b). Another accurate and time-efficient technique is supercritical fluid extraction employing carbon dioxide (Johnson and Barnett, 2003).

Given that phytoplankton communities are significant producers of lipids in the marine environment (Parrish, 1988), the study of compositional characteristics of lipid subfractions is typically associated with total lipid determination. This is because lipid subfractions serve as markers of phytoplankton community life cycles.

## 4.1 Chlorophyll pigment analysis

Multiple chlorophyll pigments are present in marine samples, with chlorophyll 'a' being the most commonly determined due to its capability to estimate total phytoplankton abundance (i.e., primary production) in seawater. Figure 3 illustrates the chemical structure of chlorophyll pigment. Various spectroscopic methods, either in absorption or fluorescence mode, are employed for its determination. A common method involves measuring spectral absorption at 665 nm in acetone solution of previously filtered samples (Lazzara et al., 1990; Dere, 1998). Alternatively, the fluorescence method with excitation at 432 nm and detection at 665 nm is less time-consuming, as it eliminates the need for filtration and pre-concentration steps owing to the high sensitivity of the fluorimetric approach (EPA Method 445.0, 1997).

A notable modification of the absorption method for chlorophyll 'a' determination has been proposed by Araujo et al. (1996). They enhance the analytical sensitivity of the method by employing multivariate calibration technique (PLSR) and utilizing all wavelengths of absorption spectra of chlorophyll between 350 nm and 700 nm.

High-Performance Liquid Chromatography (HPLC) determination of chlorophyll 'a' with spectroscopic detection is also widely utilized (Kowaleska et al., 2004; Marchand et al., 2005; van Leeuwe et al., 2006). These studies have optimized the entire procedure, including extraction and sample injection, for determining total algal pigments by HPLC.

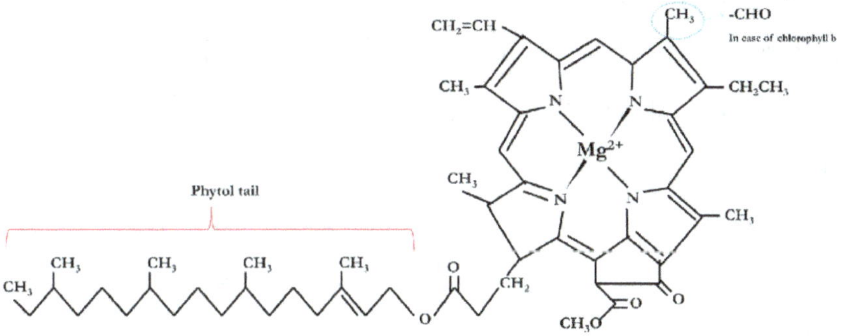

**Figure 3** Typical chemical structure of chlorophyll pigment

## 4.2 Hydrocarbon analysis

Several methods are available for the analysis of hydrocarbons in marine environments, both in seawater and sediments. Hydrocarbons in marine environments have both natural and pollutant origins (Wakeham, 1996). However, the discussion of methods to differentiate natural hydrocarbons from pollutants is beyond the scope of this paper. Therefore, this section focuses solely on methods for hydrocarbon analysis without differentiation between natural and pollutant origins.

A straightforward and rapid method for total hydrocarbon determination involves infrared spectroscopy applied to the organic extracts of seawater and sediments after a clean-up procedure that separates polar lipids from hydrocarbons (EPA

Method 418.1, 1994). This method offers high sensitivity, with a detection limit close to 0.01 mg/L or even lower attainable. Moreover, for seawater analysis, the clean-up step is often unnecessary (Chouksey et al., 2004). Infrared spectroscopy has also been used for the qualitative determination of various lipid classes, including hydrocarbons, in phytoplankton algae (Stehfest et al., 2005).

Achieving separation between saturated and aromatic hydrocarbons is a significant objective in marine studies, and various approaches exist. Wakeham (1996) outlines a comprehensive procedure for the extraction, purification, and quantification of aliphatic and polycyclic aromatic hydrocarbons in marine sediments using gas chromatography–mass spectrometry.

Thin-layer chromatography has been utilized as a separation technique between saturated and aromatic hydrocarbons before the determination of polycyclic aromatic hydrocarbons using Shploskii spectroscopy (Chernova et al., 2001). It has also been applied in marine mucilages (Mecozzi et al., 2002b) and in field experiments concerning the natural cleanup of heavy fuel oils on rocks (Jézéquel et al., 2003). Coman et al. (1997) provide detailed experimental conditions for optimizing the procedure for polycyclic aromatic hydrocarbon separation and determination by thin-layer chromatography.

Ultrasound-assisted extraction has been employed for aromatic hydrocarbons prior to their spectrofluorimetric (Rodríguez-Sanmartin et al., 2005) and gas chromatographic–mass spectroscopy (Banjoo and Nelson, 2005) determination. Lastly, surface-enhanced Raman scattering has been utilized for polycyclic aromatic hydrocarbon determination in seawater (Schmidt et al., 2004).

### 4.3 Fatty acids and sterols

Fatty acids and sterols are typically determined as ethers after derivatization using bis (trimethyl-silyl) trifluoroacetamide and detected by gas chromatography-mass spectrometry (GC-MS) (Mudge and Norris, 1997; Marchand et al., 2005; Shah et al., 2006). Alternatively, some methods involve the transmethylation of fatty acids using methanolic BF3 followed by detection with gas chromatography-flame ionization detection (GC-FID) (Pistocchi et al., 2005).

While transmethylation and silylation are commonly used derivatization methods for GC analysis, other derivatization techniques are now being applied for fatty acid analysis in complex matrices using GC and electrophoretic techniques. Rosenfeld (2002) describes derivatization with diazo reagents and bromoacetonitrile for GC analysis, as well as with pentafluorobenzyl bromide for electrophoretic analysis. The latter method is notable for its high yield achieved at room temperature and rapid reaction time. Figure 4 shows the chemical structure of fatty acid.

High-performance liquid chromatography (HPLC) is also a prevalent technique for fatty acid analysis. Compared to GC, HPLC offers the advantage of using lower temperatures during analysis, reducing the risk of unsaturated fatty acid isomerization, and enhancing the performance of fatty acid speciation analysis in complex samples. Lima and Abdalla (2002) provide a comprehensive review covering all aspects of fatty acid analysis by HPLC in biological samples, including sample preparation, identification of mobile and stationary phases, derivatization methods, and detection techniques. Specifically, their review includes a detailed section on detection techniques, encompassing UV-VIS, fluorescence, chemiluminescence, mass spectrometry, electrochemical, and light-scattering methods. Brondz (2002) offers a review of HPLC, GC, and related techniques for fatty acid analysis, citing more than 500 papers in the field.

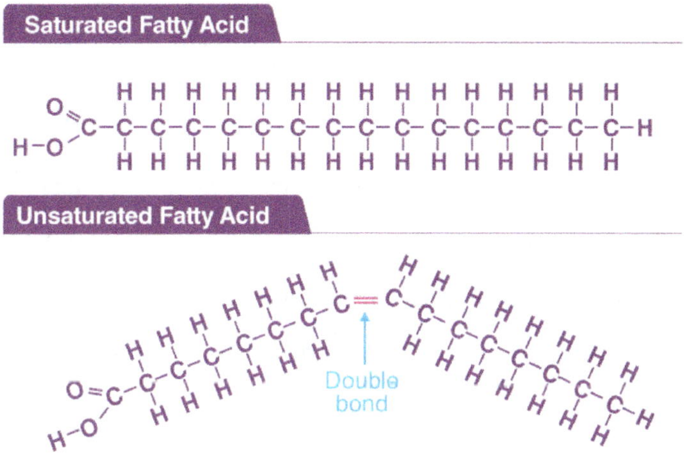

**Figure 4** Saturated and unsaturated fatty acid chemical structure

### 4.4 Phospholipids

Phospholipids are integral components of marine organisms, playing significant roles in their life cycles. Figure 5 illustrates the chemical structure of phospholipids. Qualitative determination of phospholipids in OM from sediments and algae can be achieved through Fourier transform infrared spectroscopy (Stehfest et al., 2005; Mecozzi et al., 2007) and 31P nuclear magnetic resonance spectroscopy (Kolowith et al., 2001). Suzumura and Ingall (2001) outline a fast and reliable method for the quantitative determination of lipid phosphorus using SPE combined with thin-layer chromatography-flame ionization detection (TLC-FID) analysis. An updated review on lipid phosphorus determination in marine samples, covering extraction and detection methods, is also available (Suzumura, 2005).

TLC-FID methods are increasingly adopted for their capability to separate all lipid classes without necessitating specific derivatization reactions prior to detection. These methods are employed for phospholipid determination in various marine samples, including sediments (Jézéquel et al., 2003), seawater (Volkman and Nichols, 1991), marine colloids (Liu et al., 1998), marine mucilages (Mecozzi et al., 2002b), and fish samples (Ewald et al., 1998).

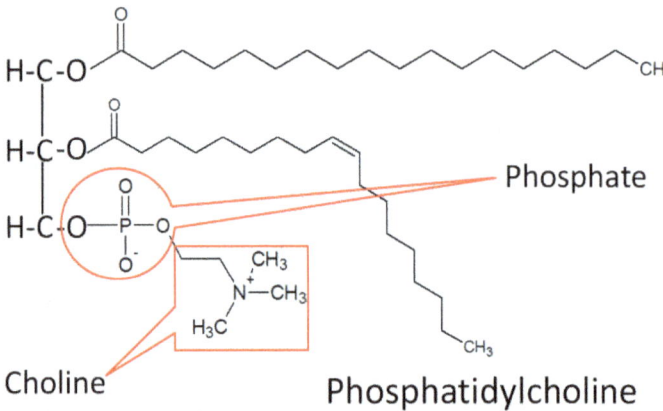

**Figure 5** Phospholipid chemical structure

## 5. Protein analysis

Proteins exhibit many analytical similarities with carbohydrates; akin to carbohydrates, proteinaceous materials are present in marine samples in a diverse range of monomeric, oligomeric, and polymeric forms.

Analytical approaches for protein determination share similarities with those for carbohydrates. Spectroscopic methods are employed for total protein analysis due to their greater sensitivity compared to the Kjeldahl method (Clesceri et al., 1989), while chromatographic methods are utilized for speciation analysis of monomeric (i.e., amino acid) composition.

An older spectrophotometric method for total protein analysis relies on sample absorption at 280 nm and 210 nm. However, this method is less sensitive and reproducible than the Lowry method (Lowry et al., 1951) and is susceptible to chemical interferences stemming from the presence of aromatic compounds in the samples (Zaia et al., 2000).

Modern spectrophotometric methods for total protein analysis are predominantly based on the reaction of $Cu^{2+}$ ions with the amino acid residues of proteins, forming a blue complex with absorption in the visible region. Both the Lowry-Hartree method (Hartree, 1972) and the biuret method (Gornall et al., 1949) exploit this specific reaction, enabling detection limits of 0.5 mg/ml and 1 mg/ml, respectively. These methods commonly employ bovine serum albumin as a calibration standard. Moreover, the biuret method has been recently modified and enhanced, achieving a detection limit of up to 20 µg/ml (Drochioiu et al., 2006).

The determination of proteins shares several analytical parallels with carbohydrates. Proteinaceous materials in marine samples encompass a broad range of monomeric, oligomeric, and polymeric compounds. Like carbohydrates, the analytical approaches for proteins involve spectroscopic methods for total protein analysis and chromatographic methods for speciation analysis of monomeric (amino acid) composition.

One commonly used spectrophotometric method for total protein analysis relies on the formation of a blue complex with the Coomassie Blue reagent, enabling a detection limit of 2 µg/ml (Bradford, 1976). This method is also applied in conjunction with microscopic counts for detecting ink-stained proteinaceous materials in marine aerosol samples (Kuznetsova et al., 2005). Another method involves bicinchoninic acid, which forms a blue complex with Cu2+ ions and proteins (Smith et al., 1985). This method has been modified to utilize RuDPCase protein as a calibration standard, deemed more representative of proteinaceous materials in marine environments than bovine serum albumin (Nguyen and Harvey, 1994). Despite their conventional nature, $Cu^{2+}$ ion complex-based methods remain prevalent in marine studies due to their simplicity and cost-effectiveness, employing common instruments like visible spectrophotometers (de Boeck et al., 2001; Lionetto et al., 2003; Mitchelmore et al., 2003; Danovaro et al., 2005; Burlando et al., 2006; Pan et al., 2006). Advancements in instrumental techniques for protein determination have emerged to overcome potential interferences encountered with $Cu^{2+}$-based methods. For instance, a spectrophotometric method based on the hydrolytic pre-treatment of proteinaceous material by 6 M HCl at 110°C for 24 h, followed by derivatization with o-phthalaldehyde and N-acetyl-L-cysteine, offers high recoveries close to 100% (Medina Hernández et al., 1990).

Gel Electrophoresis (GE) and Size Exclusion Chromatography (SEC) are emerging techniques often coupled with Matrix Assisted Laser Desorption Ionisation (MALDI). These methods provide insights into the structural aspects of proteins in samples and allow for the separation of proteins based on molecular weight (Chong et al., 2005; Powell et al., 2005). Reviews on GE, SEC, and MALDI for protein studies are available (Goetz et al., 2004; Krieg et al., 2005).

Pre-concentration treatments such as ultrafiltration enhance the analytical sensitivity for analyzing the protein fraction in POM samples (Minor et al., 2006).

The accuracy of protein determination in marine sediments is influenced by the extraction procedure, with acid hydrolysis yielding the highest recoveries (Nunn and Keil, 2006).

Liquid chromatographic techniques are employed for determining the amino acid composition of proteinaceous materials. These include HPLC methods based on the analysis of hydrolyzed amino acids derivatized with o-phthalaldehyde and 2-mercaptoethanol or ninhydrin, among others (Grasshoff et al., 1983; Touchette and Burkholder, 2001; Knicker, 2004). Some HPLC methods enable simultaneous determination of carbohydrates and amino acids without chemical derivatization (Metaxatos et al., 2003; Jandik et al., 2004).

## 6. Polyphenolic and lignin analysis

Polyphenolic compounds are generally soluble in seawater, while lignin-type materials, which have a polyphenolic origin, are insoluble (Grasshoff et al., 1983; Riemer et al., 2000; Dittmar and Kattner, 2003). Figure 6 demonstrates the chemical structure of lignin. The presence of lignin in marine organic OM samples is attributed to terrestrial inputs. Analyzing lignin in samples such as POM and sediments requires non-invasive methods or specific treatments to convert insoluble lignin into soluble compounds (Hatcher, 2004).

Solid-state $^{13}$C NMR and (Hatcher, 2004) FTIR diffuse reflectance spectroscopy (Mecozzi et al., 2001b; See and Bronk, 2005) are reliable non-invasive methods for lignin analysis. UV-VIS methods in transmission mode measure absorbance ratios at fixed wavelengths to detect aromatic compounds in marine samples (Schnitzer, 1971). These methods assess the degree of humification or evaluate the terrestrial input of OM based on absorbance ratios at specific wavelengths. However, the accuracy of UV-VIS methods for lignin estimation depends on the efficiency of reactions used to convert insoluble lignin into soluble compounds (Hatcher, 2004) (Berto et al., 2005).

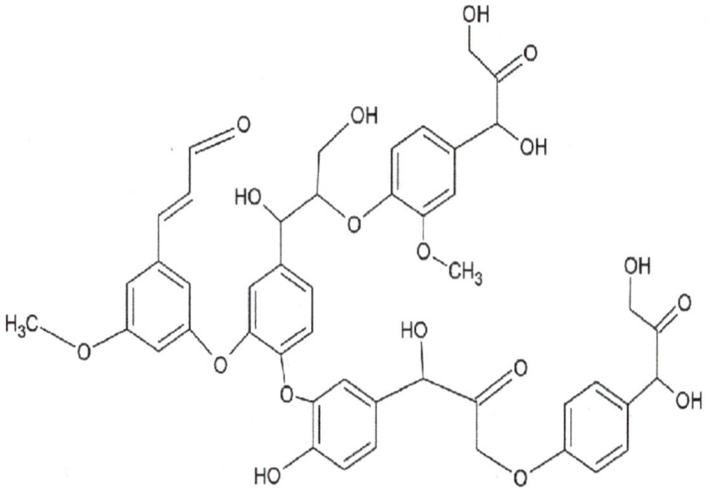

**Figure 6** Chemical structure of lignin

## 7. Structural characteristics of OM
### 7.1 OM formation and aggregation

Nowadays, studies focusing on the structural characteristics of OM and their connection to the mechanisms of formation and aggregation are increasingly common due to the significant relationships between OM and various dynamic processes in the marine environment. These studies delve into the different physicochemical forms and statuses of OM.

Within the water column and sediments, DOM exists as FAs and HAs with submicroscopic dimensions, while POM ranges from 0.2 μm to 1 μm in size (Chu-Chin et al., 1998; Verdugo, 2004; Verdugo et al., 2004). Anomalous aggregates, such as marine snow, clouds, and mucilages, have larger dimensions ranging from millimeters to meters. The solubility or insolubility of OM fractions is determined by specific structural characteristics primarily related to the presence of hydrophilic functional groups and their interactions with inorganic elements (Mingazzini and Thake, 1995) (Ishiwatari, 1992; Schulten and Schnitzer, 1995; Chu-Chin et al., 1998; Piccolo, 2001; Verdugo et al.,2004). Throughout the transition from small to large aggregate sizes, all OM fractions are interconnected through a complex equilibrium of polymerization and depolymerization reactions. These reactions lead to the transformation of functional groups and interactions between organic and inorganic fractions, influencing the overall properties and behavior of OM in the marine environment.

Various methods have been proposed to convert lignin fractions into soluble ones for spectroscopic determinations. These include converting lignin into soluble lignin sulphonates prior to fluorescent detection, oxidative methods like the CuO oxidation method, and chemolytic cleavage methods such as using tetramethylammonium hydroxide (TMAH). TMAH has the advantage of methyl-derivatizing ether-linked lignin materials, aiding in the detection of derived lignin materials.

Additionally, a spectrophotometric method assisted by multivariate calibration has been developed to determine phenolic compounds in seawater. These methods collectively contribute to understanding the presence and characteristics of lignin and polyphenolic compounds in marine environments, offering insights into organic matter dynamics and sources (Shinozuka and Lee, 1991; Wells and Goldberg, 1993; Leppard 1995, 1999; Mingazzini and Thake, 1995; Chu-Chin et al., 1998; Verdugo et al., 2004; Verdugo, 2004; Mecozzi and Pietrantonio, 2006).

## 7.2 Spectroscopic and chromatographic methods for OM structural characterization

Various spectroscopic techniques are available for characterizing the structure of OM (Mecozzi et al., 2001a). For instance, a UV spectroscopic library has been employed to assess the qualitative similarity of humic substance extracted sediments. Fluorescence spectroscopy (de Souza Sierra et al., 1994; Mingazzini et al., 1995; Ferrari et al., 1996; Stedmon et al., 2003; Boehme and Wells, 2006), FTIR spectroscopy (Kovac et al., 2002; Mecozzi and Pietrantonio, 2006), and Raman spectroscopy (Ciglenečki et al., 2000) have been utilized extensively to delve into the aggregation mechanisms of OM, elucidating the distinct roles played by carbohydrates, proteins, and lipids in their interactions with inorganic fractions such as calcium, magnesium, iron, and silicon during the transition from micro to macro aggregates.

Fluorescence spectroscopy has found widespread application in studying the dynamics of OM, including fluxes and resuspension events. Various studies have investigated OM released from coastal sediments during resuspension events, the dynamic characteristics of benthic DOM in anoxic marine pore water, and factors affecting OM dynamics in coastal sediments (Chen et al., 1993). Excitation–emission matrix spectroscopy, used by Coble, has been applied to characterize marine and terrestrial DOM (Everett et al., 1999), including humic substances extracted from corals for use as paleoenvironmental indicators. Combining absorption and emission spectroscopy has enabled researchers to examine the role of nitrogen-containing compounds in OM formation and assess the contribution of riverine OM in specific environments (Matthews et al., 1996).

Fluorescence spectroscopy, in conjunction with potentiometric techniques (Minor et al., 2002; Hutta and Gora, 2003; Li et al., 2003); has facilitated the characterization of the acid–base properties of humic substances. This combination involves conventional acid–base titration of humic substances coupled with synchronous spectra analysis (de la Rosa et al., 2006), followed by the application of chemometric techniques such as PCA and Evolving Factor Analysis (EFA) to determine the pKa values of functional groups present (Biers et al., 2007).

Nuclear magnetic resonance (NMR) spectroscopy (Knicker, 2004), including $^1$H and $^{13}$C NMR, has provided insights into the structural behavior of mucilaginous macro aggregates of OM and the formation and stabilization of refractory nitrogen compounds in marine sediments (Kovac et al., 2002). Tandem mass spectrometry, electron scanning microscopy, and X-ray diffraction have also been employed for studying polar interactions and determining colloid number and size distribution in particulate OM (Chen et al., 2007).

Size-exclusion chromatography (SEC) (Esteves da Silva and Machado, 1995), which separates macromolecular complexes based on their hydrodynamic volume (Dittmar and Kattner, 2003), has been widely used in combination with various detection methods such as UV detection (Hertkorn et al., 2006), mass spectrometry (Powell et al., 2005), and tangential flow ultrafiltration to study the molecular weight of humic fractions (Chu-Chin et al., 1998) and other recalcitrant OM components (Brown et al., 2004) (Wells and Goldberg, 1991).

## 8. Conclusion

This chapter systematically evaluates methodologies employed in the analysis of natural marine organic matter (OM), encompassing various chemical constituents such as carbohydrates, proteins, lipids, and lignin, with a focus on analytical precision, accuracy, and time efficiency. Furthermore, diverse techniques for sample preparation and structural characterization of OM are delineated and analyzed. It is posited that comprehensive research endeavors, such as those pertaining to the analysis of marine OM, necessitate a profound understanding of inherent analytical challenges, including but not limited to issues of precision and accuracy, which are meticulously addressed within this review. Moreover, it is emphasized that the realm of OM analysis and characterization demands perpetual advancement in analytical methodologies to meet the evolving demands of environmental research, ensuring the provision of dependable insights and solutions.

**References**

Almgren, T., Josefsson, B. and Nyquist, G. (1975) 'A fluorescence method for studies of spent sulfite liquor and humic substances in sea water', *Analytica Chimica Acta*, Vol. 78, No. 2, pp.411–422.

AOAC (1996) *Official Methods of Analysis of the Official Association of Analytical Chemists*,
16th ed., Supplement, The Association Arlington, VA, USA.

Araujo, P.W., Cirovic, D.A. and Brereton, R.G. (1996) 'Multivariate calibration of chlorophyll a using partial least squares and electronic absorption spectroscopy', *Analyst*, Vol. 121, No. 5, pp.581–584.

Banjoo, D.R. and Nelson, P.K. (2005) 'Improved ultrasonic extraction procedure for the determination of polycyclic aromatic hydrocarbons in sediments', *Journal of Chromatography A*, Vol. 1066, Nos. 1–2, pp.9–18.

Belcher, R., Nutten, A.J. and Sambrook, C.M. (1954) 'The determination of glucosamine', *Analyst*, Vol. 79, pp.201–208.

Belzile, N., Joly, H.A. and Li, H. (1997) 'Characterization of humic substances extracted from Canadian lake sediments', *Canadian Journal of Chemistry*, Vol. 75, No. 1, pp.14–27.

Benner, R., Biddanda, B., Black, B. and McCarthy, M. (1997) 'Abundance, size distribution, and stable carbon and nitrogen isotopic compositions of marine organic matter isolated by tangential-flow ultrafiltration', *Marine Chemistry*, Vol. 57, Nos. 3–4, pp.243–263.

Beramendi-Orosco, L., Snape, C.E. and Large, D.J. (2006) 'Stable carbon isotope analysis of wood hydropyrolysis residues: a potential indicator for the extent of cross-linking between lignin and polysaccharides', *Organic Geochemistry*, Vol. 37, No. 1, pp.64–71.

Berto, D., Giani, M., Taddei, P. and Bottura, G. (2005) 'Spectroscopic evidence of the marine origin of mucilages in the Northern Adriatic Sea', *Science of the Total Environment*, Vol. 353, Nos. 1–3, pp.247–257.

Biers, E.J., Zepp, R.G. and Moran, M.A. (2007) 'The role of nitrogen in chromophoric and fluorescent dissolved organic matter formation', *Marine Chemistry*, Vol. 103, Nos. 1–2, pp.46–60.

Bligh, E.G. and Dyer, W.J. (1959) 'A rapid method of total lipid extraction and purification',
*Canadian Journal of Biochemistry and Physiology*, Vol. 37, p.911.

Blumenkrantz, N. and Asboe-Hansen, G. (1973) 'A quick and specific assay for hydroxyproline',
*Analytical Biochemistry*, Vol. 55, No. 1, pp.288–291.

Boehme, J. and Wells, M. (2006) 'Fluorescence variability of marine and terrestrial colloids: examining size fractions of chromophoric dissolved organic matter in the damariscotta river estuary', *Marine Chemistry*, Vol. 101, Nos. 1–2, pp.95–103.

Borch, N.H. and Kirchman, D.L. (1997) 'Concentration and composition of dissolved combined neutral sugars (polysaccharides) in seawater determined by HPLC-PAD', *Marine Chemistry*, Vol. 57, Nos. 1–2, pp.85–95.

Bradford, M.M. (1976) 'A rapid and sensitive method for the quantitation of microgram quantities of protein utilizing the principle of protein dye binding', *Analytical Biochemistry*, Vol. 72, Nos. 1–2, pp.248–254.

Brondz, I. (2002) 'Development of fatty acid analysis by high-performance liquid chromatography, gas chromatography, and related techniques', *Analytica Chimica Acta*, Vol. 465, Nos. 1–2, pp.1–37.

Brown, A., McKnight, D.M., Chin, Y., Roberts, E.C. and Uhle, M. (2004) 'Chemical characterization of dissolved organic material in Pony Lake, a saline coastal pond in Antarctica', *Marine Chemistry*, Vol. 89, Nos. 1–4, pp.327–337.

Burdige, D.J., Kline, S.W. and Chen, W. (2004) 'Fluorescent dissolved organic matter in marine sediment pore waters', *Marine Chemistry*, Vol. 89, Nos. 1–4, pp.289–311.

Burlando, B., Berti, E. and Viarengo, A. (2006) 'Effects of seawater pollutants on protein tyrosine phosphorylation in mussel tissues', Aquatic Toxicology, Vol. 78S, pp.S79–S85.

Butcher, E.C. and Lowry, O.H. (1976) 'Measurement of nanogram quantities of protein by hydrolysis followed by reaction with orthophthalaldehyde or determination of glutamate', Analytical Biochemistry, Vol. 76, No. 2, pp.502–523.

Calace, N., Cardellicchio, N., Petronio, B.M., Pietrantonio, M. and Pietroletti, M. (2006a)

'Sedimentary humic substances in the northern Adriatic sea (Mediterranean sea)', Marine Environmental Research, Vol. 61, No. 1, pp.40–58.

Calace, N., Palmieri, N., Mirante, S., Petronio, B.M. and Pietroletti, M. (2006b) 'Dissolved and particulate humic substances in water channels in the historic centre of Venice', Water Research, Vol. 40, No. 6, pp.1109–1118.

Chen, R.F., Bada, J.L. and Suzuki, Y. (1993) 'The relationship between dissolved organic carbon (DOC) and fluorescence in anoxic marine porewaters: implications for estimating benthic DOC fluxes', Geochimica et Cosmochimica Acta, Vol. 57, No. 9, pp.2149–2153.

Chen, Z., Hu, C., Conmy, R.N., Muller-Karger, F. and Swarzenski, P. (2007) 'Colored dissolved organic matter in Tampa Bay, Florida', Marine Chemistry, Vol. 104, Nos.1–2, pp.98–109.

Chernova, T.G., Rao, P.S., Pikovskii, Yu.I., Alekseeva, T.A., Nagender Nath, B., Ramalingeswara Rao, B. and Rao, Ch.M. (2001) 'The composition and the source of hydrocarbons in sediments taken from the tectonically active Andaman Backarc Basin, Indian Ocean', Marine Chemistry, Vol. 75, Nos. 1–2, pp.1–15.

Chong, K., Tham, S.Y., Foo, J., Lam, T.J. and Chong, A. (2005) 'Characterisation of proteins in epidermal mucus of discus fish (Symphysodon spp.) during parental phas', Aquaculture, Vol. 249, Nos. 1–4, pp.469–476.

Chouksey, M.K., Kadam, A.N. and Zingde, M.D. (2004) 'Petroleum hydrocarbon residues in the marine environment of Bassein-Mumbai', Marine Pollution Bulletin, Vol. 49, Nos. 7–8, pp.637–647.

Chu-Chin, W., Orellana, M.V. and Verdugo, P. (1998) 'Spontaneous assembly of marine dissolved organic matter into polymer gels', Nature, Vol. 391, No. 6667, pp.568–572.

Ciglenečki, I., Ćosović, B., Vojvodić, V., Plavšić, M., Furić, K., Minacci, A. and Baldi, F. (2000)
'The role of reduced sulfur species in the coalescence of polysaccharides in the Adriatic Sea',
Marine Chemistry, Vol. 71, Nos. 3–4, pp.233–249.

Clesceri, L.S., Greenberg, A.E., Rhodes Trussel, R. (1989) Standard Method for the Examination of Water and Wastewater, American Public Health Association, 17th ed., USA.

Coble, P.G. (1996) 'Characterization of marine and terrestrial DOM in seawater using excitation-emission matrix spectroscopy', Marine Chemistry, Vol. 51, No. 4, pp.325–346.

Coman, V., Măruţoiu, C. and Puiu, S. (1997) 'Optimization of separation of some polycyclic aromatic compounds by thin-layer chromatography', Journal of Chromatography A, Vol. 779, Nos. 1–2, pp.321–328.

Danovaro, R., Armeni, M., Luna, G.M., Corinaldesi, C., Dell'Anno, A., Ferrari, C., Fiordelmondo, C., Gambi, C., Gismondi, M., Manini, E., Mecozzi, M., Perrone, F.M., Pusceddu, A. and Giani, M. (2005) 'Exo-enzymatic activities and dissolved organic pools in relation with mucilage development in the Northern Adriatic Sea', Science of the Total Environment, Vol. 353, Nos. 1–3, pp.189–203.

de Boeck, G., Grosell, M. and Wood, C. (2001) 'Sensitivity of the spiny dogfish (Squalus acanthias) to waterborne silver exposure', Aquatic Toxicology, Vol. 54, Nos. 3–4, pp.261–275.

de la Rosa, J.M., González-Pérez, J.A., González-Vázquez, R., Verdejo, T., de Andrés, J.R., Sánchez-García, L., Hatcher, P.G., Knicker, H., Terán, A., Polvillo, O. and González-Vila, F.J. (2006) 'Detection of nitrogen stable forms in marine sediments by high resolution magic angle spinning (HRMAS) 1H nuclear magnetic resonance (NMR)', Geophysical Research Abstracts, Vol. 8, p.3925.

De Souza Sierra, M.M., Donard, O.F.X., Lamotte, M., Belin, C. and Ewald, M. (1994) 'Fluorescence spectroscopy of coastal and marine waters', *Marine Chemistry*, Vol. 47, No. 2, pp.127–144.

Dere Ş., Güneş T. and Sıvacı R. (1998) 'Spectrophotometric determination of chlorophyll- A,B and total carotenoid contents of some algae species using different solvents', *Turkish Journal of Botany*, Vol. 22, pp.13–18.

Di Nezio, M., Pistonesi, M.F., Franoso, W. D., Pontes, M.J.C., Goicoecea, H.G. and Araujo, M.C.U. (2007) 'Successive projections algorithm improving the multivariate simultaneous direct spectrophotometric determination of five phenolic compounds in sea water', *Microchemical Journal*, Vol. 85, pp.194–200.

Dittmar, T. and Kattner, G. (2003) 'Recalcitrant dissolved organic matter in the ocean: major contribution of small amphiphilics', *Marine Chemistry*, Vol. 82, Nos.1–2, pp.115–123.

Doner, L.W. (2001) 'Determining sugar composition of food gum polysaccharides by HPTLC', *Chromatographia*, Vol. 53, Nos. 9–10, pp.579–581.

Drochioiu, G., Damoc, N.E. and Przybylky, M. (2006) 'Novel UV assay for protein determination and the characterization of copper-protein complexes by mass spectrometry', *Talanta*, Vol. 69, No. 3, pp.556–564.

Dubois, M., Gilles, K.A., Hamilton, J.K., Rebers, P.A. and Smith, F. (1956) 'Colorimetric method for determination of sugars and related substances', *Analytical Chemistry*, Vol. 28, pp.350–356.

EPA Method 418.1 (1994) *Test Methods for Petroleum Hydrocarbon Analysis*, Cincinnati USA. EPA Method 445.0 (1997) *In vitro Determination of Chloropyll a and Pheophytin in Marine and Freshwater Algae by Fluorescence*, Cincinnati USA.

Esteves da Silva, J.C.G. and Machado, A.A.S.C. (1995) *Analytical Letters*, Vol. 28, No. 13, pp.2401–2411.

Esteves, V.I., Cordeiro, N.M.A. and da Costa Duarte, A. (1995) 'Variation on the adsorption efficiency of humic substances from estuarine waters using XAD resins', *Marine Chemistry*, Vol. 51, No. 1, pp.61–66.

Everett, R.E., Chin Yu-Ping, Aiken G.R. (1999) 'High-pressure size exclusion chromatography analysis of dissolved organic matter isolated by tangential-flow ultrafiltration', *Limnology and Oceanography*, Vol. 44, No. 5, pp.1316–1322.

Ewald, G., Bremle, G. and Karlsonn, A. (1998) 'Differences between bligh and dyer and soxhlet extractions of PCBs and lipids from fat and lean fish muscle: implications for data evaluation', *Marine Pollution Bulletin*, Vol. 36, No. 3, pp.222–230.

Ferrari, G.M., Dowell, M.D., Grossi, S. and Targa, C. (1996) 'Relationship between the optical properties of chromophoric dissolved organic matter and total concentration of dissolved organic carbon in the southern Baltic Sea region', *Marine Chemistry*, Vol. 55, Nos. 3–4, pp.299–316.

Filisetti-Cozzi, T.M.C.C. and Carpita, N.C. (1991) 'Measurement of uronic acids without interference from neutral sugars', *Analytical Biochemistry*, Vol. 197, No. 1, pp.157–162.

Filley, T.R., Hatcher, P.G., Shortle, W.C. and Presuth, R.T. (2000) 'The application of 13C-labeled tetramethylammonium hydroxide (13C-TMAH) thermochemolysis to the study of fungal degradation of wood', *Organic Geochemistry*, Vol. 31, Nos. 2–3, pp.181–198.

Folch, J., Lees, M. and Stanley, G.H. (1957) 'A simple method for the isolation and purification of total lipids from animal tissues', *The Journal of Biological Chemistry*, Vol. 226, p.497.

Fooken, U. and Liebezeit, G. (2000) 'Distinction of marine and terrestrial origin of humic acids in North Sea surface sediments by absorption spectroscopy', *Marine Geology*, Vol. 164, Nos. 3–4, pp.173–181.

Garcia, D., Cegarra J., Bernal, M.P. and Navarro, A. (1993) 'Communications i soil,
*Science and Plant Analysis*, Vol. 24, Nos. 13–14, pp.1481–1494.

Giani, M., Berto, D., Zangrando, V., Castelli, S., Sist, P. and Urbani, R. (2005) 'Chemical characterization of different typologies of mucilaginous aggregates in the northern Adriatic Sea', *Science of the Total Environment*, Vol. 353, Nos. 1–3, pp.232–246.

Goetz, H., Kuschel, M., Wulff, T., Sauber, C., Miller, C., Fisher, S. and Woodward, C. (2004) 'Comparison of selected analytical techniques for protein sizing, quantitation and molecular weight determination', *Journal of Biochemical and Biophysical Methods*, Vol. 60, No. 3, pp.281–293.

Gornall, A.G., Bardawill, C.J. and David, M.M. (1949) 'Determination of serum proteins by means of the biuret reaction', *Journal of Biological Chemistry*, Vol. 177, pp.751–765.

Grasshoff, K., Erhardt, M. and Kremling, K. (Eds.) (1983) *Methods of Seawater Analysis*, Verlag Chemie, Weinheim, Germany, p.600.

Guo, L. and Santschi, P.H. (1996) 'A critical evaluation of the cross-flow ultrafiltration technique for sampling colloidal organic carbon in seawater', *Marine Chemistry*, Vol. 55, Nos. 1–2, pp.113–127.

Guo, L., Wen, L., Tang, D. and Santschi, P.H. (2000) 'Re-examination of cross-flow ultrafiltration for sampling aquatic colloids: evidence from molecular probes', *Marine Chemistry*, Vol. 69, Nos. 1–2, pp.75–90.

Han, N.S. and Robyt, J.F. (1998) 'The mechanism of Acetobacter xylinum cellulose biosynthesis: direction of chain elongation and the role of lipid pyrophosphate intermediates in the cell membrane', *Carbohydrate Research*, Vol. 313, No. 2, pp.125–133.

Hartree, E.F. (1972) 'Determination of protein: a modification of the Lowry method that gives a linear photometric response', *Analytical Biochemistry*, Vol. 48, pp.422–427.

Hatcher, P.G. (2004) 'The CHNs of organic geochemistry: characterization of molecularly uncharacterized non-living organic matter', *Marine Chemistry*, Vol. 92, Nos. 1–4, pp.5–8.

Hedges, J.I., Eglinton, G., Kirchman, D.L., Arnosti, C., Derenne, S., Evershed, R.P., Kögel-Knaber, I., de Leew, J.W., Littke, R., Michaelis, W. and Rullköter, J. (2000) 'The molecularly-uncharacterized component of nonliving organic matter in natural environments', *Organic Geochemistry*, Vol. 31, No. 10, pp.945–958.

Hertkorn, N., Benner, R., Frommberger, M., Schmitt-Kopplin, P., Witt, M., Kaiser, K., Kettrup, A. and Hedges, J.I. (2006)

'Characterization of a major refractory component of marine dissolved organic matter', *Geochimica et Cosmochimica Acta*, Vol. 70, No. 12, pp.2990–3010.

Hung, C. and Santschi, P. (2001) 'Spectrophotometric determination of total uronic acids in seawater using cation-exchange separation and pre-concentration by lyophilization', *Analytica Chimica Acta*, Vol. 427, No. 1, pp.111–117.

Hutta, M. and Gora, R. (2003) 'Novel stepwise gradient reversed-phase liquid chromatography separations of humic substances, air particulate humic-like substances and lignins', *Journal of Chromatography A*, Vol. 1012, No. 1, pp.67–79.

Ishiwatari, R. (1992) 'Macromolecular material (humic substance) in the water column and sediments', *Marine Chemistry*, Vol. 39, Nos. 1–3, pp.151–166.

Jandik, P., Cheng, J. and Avdalovic, N. (2004) 'Analysis of amino acid-carbohydrate mixtures by anion exchange chromatography and integrated pulsed amperometric detection', *Journal of Biochemical and Biophysical Methods*, Vol. 60, No. 3, pp.191–203.

Jézéquel, R., Menot, L., Merlin, F.X. and Price, R.C. (2003) 'Natural cleanup of heavy fuel oil on rocks: an in situ experiment', *Marine Pollution Bulletin*, Vol. 46, No. 8, pp.983–990.

Johnson, K. and Sieburth, J.McN. (1977) 'Dissolved carbohydrates in seawater. 1. A precise spectrophotometric analysis for monosaccharides', *Marine Chemistry*, Vol. 5, pp.1–13.

Johnson, R.B. and Barnett, H.J. (2003) 'Determination of fat content in fish feed by supercritical fluid extraction and subsequent lipid classification of extract by thin layer chromatography-flame ionization detection', *Aquaculture*, Vol. 216, Nos. 1–4, pp.263–282.

Josefsson, B., Uppström, L. and Östling, G. (1972) 'Automatic spectrophotometric procedures for the determination of the total amount of dissolved carbohydrates in sea water', *Deep-Sea Research*, Vol. 19, pp.385–395.

Kaiser, K. and Benner, R. (2000) 'Determination of amino sugars in environmental samples with high salt content by high-performance anion-exchange chromatography and pulsed amperometric detection', *Analytical Chemistry*, Vol. 72, No. 11, pp.2566–2572.

Knicker, H. (2004) 'Stabilization of N-compounds in soil and organic-matter-rich sediments – What is the difference?', *Marine Chemistry*, Vol. 92, Nos. 1–4, pp.167–195.

Koh, T.S. (1983) 'Ultrasonic preparation of fat-free biological materials for elemental analysis',

*Analytical Chemistry*, Vol. 55, No. 11, pp.1814, 1815.

Kolowith, L.C., Ingall, E.D. and Benner, R. (2001) 'Composition and cycling of marine organic phosphorus', *Limnology and Oceanography*, Vol. 46, No. 2, pp.309–320.

Komada, T., Schofield, O.M.E. and Reimers, C.E. (2002) 'Fluorescence characteristics of organic matter released from coastal sediments during resuspension', *Marine Chemistry*, Vol. 79, No. 2, pp.81–97.

Komada, T., Reimers, C.E., Luther, G.W. and Burdige, D.J. (2004) 'Factors affecting dissolved organic matter dynamics in mixed-redox to anoxic coastal sediments', *Geochimica et Cosmochimica Acta*, Vol. 68, No. 20, pp.4099–4111.

Kovac, N., Bajt, O., Faganeli, J., Sket, B. and Orel B. (2002) 'Study of macroaggregate composition using FT-IR and 1H-NMR spectroscopy', *Marine Chemistry*, Vol. 78, No. 4, pp.205–215.

Kowaleska, G., Wawrzyniak-Wydrowska, B. and Szymczak-Zyla, M. (2004) 'Chlorophyll a and its derivatives in sediments of the Odra estuary as a measure of its eutrophication', *Marine Pollution Bulletin*, Vol. 49, No. 3, pp.148–153.

Krieg, R.C., Dong, Y., Schwamborn, K. and Knuechel, R. (2005) 'Protein quantification and its tolerance for different interfering reagents using the BCA-method with regard to 2D SDS PAGE', *Journal of Biochemical and Biophysical Methods*, Vol. 65, No. 1, pp.13–19.

Kuznetsova, M., Lee, C. and Aller, J. (2005) 'Characterization of the proteinaceous matter in marine aerosols', *Marine Chemistry*, Vol. 96, Nos. 3–4, pp.359–377.

Lara, R.J. and Thomas, D.N. (1994). 'XAD-fractionation of 'new' dissolved organic matter: Is the hydrophobic fraction seriously underestimated?', *Marine Chemistry*, Vol. 47, No. 1, pp.93–96.

Lazzara, L., Bianchi, F., Falcucci, M., Hull, V., Modigh, M. and Ribera d'Alcalà, M. (1990)
'Pigmenti clorofilliani', in Thalassia, N. (Ed.): *Metodi Nell'ecologia del Plancton Marino*, Vol. 11, pp.207–223.

Ledl, F. and Schleicher, E. (1990) 'New aspects of the Maillard reaction in foods and in the human body', *Angewandte Chemie (International Edition in English)*, Vol. 29, No. 6, pp.565–594.

Leppard, G.G. (1995) 'The characterization of algal and microbial mucilages and their aggregates in aquatic ecosystems', *Science of the Total Environment*, Vol. 165, pp.103–131.

Leppard, G.G. (1999) 'Structure/function/activity relationships in marine snow. Current understanding and suggested research thrusts', *Annali dell'Istituto Superiore di Sanità*, Vol. 35, No. 3, pp.389–395.

Li, F., Yuasa, A., Chiharada, H. and Matsui, Y. (2003) 'Polydisperse adsorbability composition of several natural and synthetic organic matrices', *Journal of Colloid and Interface Science*, Vol. 265, No. 2, pp.265–275.

Lima, E.S. and Abdalla, D.S.P. (2002) 'High-performance liquid chromatography of fatty acids in biological samples', *Analitica Chimica Acta*, Vol. 465, pp.81–91.

Lin, J. and McKeon, T.A. (2005) in Jiann-Tsyh (Ken) (Ed.): *HPLC of Acyl Lipids*, HNB Publ.,

– X, 576 S.: Ill., Graph, Darst, New York, NY, ISBN 0-9728061-1-3.

Lionetto, M.G., Caricato, R., Giordano, M.E., Pascariello, M.F., Marinosci, L. and Schettino, T. (2003) 'Integrated use of biomarkers (acetylcholinesterase and antioxidant enzymes activities) in Mytilus galloprovincialis and Mullus barbatus in an Italian coastal marine area', *Marine Pollution Bulletin*, Vol. 46, No. 3, pp.324–330.

Liška, I. (2000) 'Fifty years of solid-phase extraction in water analysis – historical development and overview', *Journal of Chromatography A*, Vol. 885, Nos. 1–2, pp.3–16.

Liu, Q., Parrish, C.C. and Helleur, R. (1998) 'Lipid class and carbohydrate concentrations in marine colloids', *Marine Chemistry*, Vol. 60, Nos. 3–4, pp.177–188.

Lowry, O.H., Rosenbrough, N.J., Farr, A.L. and Randall, R.J. (1951) 'Protein measurement with the Folin phenol reagent', *Journal of Biological Chemistry*, Vol. 193, pp.265–275.

Marchand, D., Marty, J., Miquel, J. and Rontani J. (2005) 'Lipids and their oxidation products as biomarkers for carbon cycling in the northwestern Mediterranean Sea: results from a sediment trap study', *Marine Chemistry*, Vol. 95, Nos. 1–2, pp.129–147.

Marsit, C.J., Fried, B. and Sherma, J.J. (2000) 'Carbohydrate analysis, by high performance thin layer chromatography, of Cerithidea californica (Gastropoda: Prosobranchia)', *Journal of Liquid Chromatography and Related Technologies*, Vol. 23, No. 15, pp.2413–2417.

Matthews, B.J.H., Jones, A.C., Theodoru, N.K. and Tudhope, A.W. (1996) 'Excitation-emission-matrix fluorescence spectroscopy applied to humic acid bands in coral reefs', *Marine Chemistry*, Vol. 55, Nos. 3–4, pp.317–332.

McCarthy, M., Hedges, J.I. and Benner, R. (1996) 'Major biochemical composition of dissolved high molecular weight organic matter in seawater', *Marine Chemistry*, Vol. 55, Nos. 3–4, pp.281–297.

Mecozzi, M., Pietrantonio, E. and Amici, M. (1998) 'Qualitative and quantitative studies of humic acid and fulvic acid fraction of humic substance in marine sediments by ultraviolet spectrocopy', *Fresenius Environmental Bulletin*, Vol. 7, pp.605–614.

Mecozzi, M., Pietrantonio, E., Amici, M. and Acquistucci, R. (1999) 'Ultrasound-assisted analysis of total carbohydrates in environmental and food samples', *Ultrasonics Sonochemistry*, Vol. 6, No. 3, pp.133–139.

Mecozzi, M., Dragone, P., Amici, M. and Pietrantonio, E. (2000) 'Ultrasound assisted extraction and determination of the carbohydrate fraction in marine sediments', *Organic Geochemistry*, Vol. 31, No. 12, pp.1797–1803.

Mecozzi, M., Cardarilli, D., Pietrantonio, E. and Amici, M. (2001a) 'Estimation of similarity in the qualitative composition of humic substance in marine sediments by means of an UV spectroscopic library', *Chemistry and Ecology*, Vol. 17, pp.239–254.

Mecozzi, M.R., Acquistucci, R., Di Noto, V., Pietrantonio, E., Amici, M. and Cardarilli, D. (2001b)
'Characterization of mucilage aggregates in Adriatic and Tyrrhenian Sea: structure similarities between mucilage

samples and the insoluble fractions of marine humic substance',
*Chemosphere*, Vol. 44, No. 4, pp.709–720.

Mecozzi, M., Acquistucci, R., Amici, M. and Cardarilli, D. (2002a) 'Improvement of an ultrasound assisted method for the analysis of total carbohydrate in environmental and food samples', *Ultrasonics Sonochemistry*, Vol. 9, No. 4, pp.219–223.

Mecozzi, M., Amici, A., Romanelli, G., Pietrantonio, E. and Deluca, A. (2002b) 'Ultrasound extraction and thin layer chromatography-flame ionization detection analysis of the lipid fraction in marine mucilage samples', *Journal of Chromatography A*, Vol. 963, Nos. 1–2, pp.363–373.

Mecozzi, M., Pietrantonio, E., Amici, M. and Romanelli, G. (2002c) 'An ultrasound assisted extraction of the available humic substance from marine sediments', *Ultrasonics Sonochemistry*, Vol. 9, No. 1, pp.11–18.

Mecozzi, M. and Pápai, Z.J. (2004) 'Application of curve fitting in thin-layer chromatography-flame ionization detection analysis of the carbohydrate fraction in marine mucilage and marine snow samples from Italian Seas', *Journal of Chromatographic Science*, Vol. 42, No. 5, pp.268–274

Mecozzi, M. (2005) 'Estimation of total carbohydrate amount in environmental samples by the phenol-sulphuric acid method assisted by multivariate calibration', *Chemometrics and Intelligent Laboratory Systems*, Vol. 79, Nos. 1–2, pp.84–90.

Mecozzi, M., Pietrantonio, E., Di Noto, V. and Pápai, Z. (2005) 'The humin structure of mucilage aggregates in the Adriatic and Tyrrhenian seas: hypothesis about the reasonable causes of mucilage formation', *Marine Chemistry*, Vol. 95, Nos. 3–4, pp.255–269.

Mecozzi, M. and Pietrantonio, E. (2006) 'Carbohydrates proteins and lipids in fulvic and humic acids of sediments and its relationships with mucilaginous aggregates in the Italian seas',
*Marine Chemistry*, Vol. 101, Nos. 1–2, pp.27–39.

Mecozzi, M., Onorati, F., Oteri, F. and Sarni, A. (2007) 'Characterisation of a bioassay using the marine alga dunaliella tertiolecta associated with spectroscopic (visible and infrared) detection', *International Journal of Environment and Pollution*, in press.

Medina Hernández, M.J., Villanueva Camañas, R.M., Monfort Cuenca, E. and Garcia Alvarez-Coque, M.C. (1990) 'Determination of the protein and free amino acid content in a sample using o-phthalaldehyde and N-acetyl-L-cysteine', *Analyst*, Vol. 115, No. 8, pp.1125–1128.

Metaxatos, A., Panagiotopolous, C. and Ignatiedes L. (2003) 'Monosaccharide and aminoacid composition of mucilage material produced from a mixture of four phytoplanktonic taxa', *Journal of Experimental Marine Biology and Ecology*, Vol. 294, No. 2, pp.203–217.

Mingazzini, M., Colombo, S. and Ferrari, G.M. (1995) 'Application of spectrofluorimetric techniques to the study of marine mucilages in the Adriatic Sea: preliminary results', *Science of the Total Environment*, Vol. 165, pp.133–144.

Mingazzini, M. and Thake, B.M. (1995) 'Summary and conclusions of the workshop on marine mucilages in the Adriatic Sea and elsewhere', *Science of the Total Environment*, Vol. 165, pp.9–14.

Minor, E.C., Simjouw, J-P., Boonc, J.J., Kerkhoff, A.E. and van der Horst, J. (2002) 'Estuarine/marine UDOM as characterized by size-exclusion chromatography and organic mass spectrometry', *Marine Chemistry*, Vol. 78, Nos. 2–3, pp.75–102.

Minor, E.C., Simjouw, J-P. and Mulholland, M.R. (2006) 'Seasonal variations in dissolved organic carbon concentrations and characteristics in a shallow coastal bay', *Marine Chemistry*, Vol. 101, Nos. 3–4, pp.166–179.

Mitchelmore, C.L., Verde, E.A., Ringwood, A.H. and Weis, V.M. (2003) 'Differential accumulation of heavy metals in the sea anemone Anthopleura elegantissima as a function of symbiotic state', *Aquatic Toxicology*, Vol. 64, No. 3, pp.317–329.

Moreda-Piñeiro, A., Bermejo-Barrera, A. and Bermejo-Barrera, P. (2004) 'New trends involving the use of ultrasound energy for the extraction of humic substances from marine sediments', *Analytica Chimica Acta*, Vol. 524, Nos. 1–2, pp.97–107.

Moreda-Piñeiro, A., Seco-Gesto, E.M., Bermejo-Barrera, A. and Bermejo-Barrera, P. (2006) 'Characterization of surface marine sediments from Ría de Arousa estuary according to extractable humic matter content', *Chemosphere*, Vol. 64, No. 5, pp.866–873.

Mudge, S.M. and Norris, C.E. (1997) 'Lipid biomarkers in the Conwy Estuary (North Wales, U.K.): a comparison between fatty alcohols and sterols', *Marine Chemistry*, Vol. 57, Nos. 1–2, pp.61–84.

Myklestad, S.M., Skånøy, E. and Hestmann, S. (1997) 'A sensitive and rapid method for analysis of dissolved mono- and polysaccharides in seawater', *Marine Chemistry*, Vol. 56, Nos. 3–4, pp.279–286.

Nguyen, R.T. and Harvey, H.R. (1994) 'A rapid micro-scale method for the extraction and analysis of protein in marine sample', *Marine Chemistry*, Vol. 45, Nos. 1–2, pp.1–14.

Nunn, B.L. and Keil, R.G. (2006) 'A comparison of non-hydrolytic methods for extracting amino acids and proteins from coastal marine sediments', *Marine Chemistry*, Vol. 98, No. 1, pp.31–42.

Oades, M.J. (1967) 'Gas-liquid chromatography of alditol acetates and its application to the analysis of sugars in complex hydrolysates', *Journal of Chromatography A*, Vol. 28, No. 2, pp.246–252.

Pakulski, J.D. and Benner, R. (1992) 'An improved method for the hydrolysis and MBTH analysis of dissolved and particulate carbohydrates in seawater', *Marine Chemistry*, Vol. 40, Nos. 3–4, pp.143–160.

Pan, L.Q., Ren, J. and Liu, J. (2006) 'Responses of antioxidant systems and LPO level to benzo(a)pyrene and benzo(k)fluoranthene in the haemolymph of the scallop Chlamys ferrari',
*Environmental Pollution*, Vol. 141, No. 3, pp.443–451.

Parrish, C.C. (1988) 'Dissolved and particulate marine lipid classes: a review', *Marine Chemistry*, Vol. 23, Nos. 1–2, pp.17–40.

Passow, U. and Alldredge, A.L. (1995) 'A dye-binding assay for the spectrophotometric measurement of transparent exopolymer particles (TEP)', *Limnology and Oceanography*, Vol. 40, No. 7, pp.1326–1335.

Petronio, B.M. (1997) 'Techniques of extraction and analytical methods for humic substances in sea water and sediments', *Water Science and Technology Library*, Vol. 25, pp.211–224.

Piccolo, A. (2001) 'The supramolecular structure of humic substances', *Soil Science*, Vol. 166, No. 11, pp.810–832.

Pietrantonio, E., Amici, M. and Mecozzi, M. (2003) 'Comparison of purification characteristics of fulvic acids from marine sediments using various amberlite, amberchrom and supelite resins', *Chromatographia*, Vol. 57, pp.S137–S141.

Pistocchi, R., Trigari, G., Serrazanetti, G.P., Taddei, P., Monti, G., Palamidesi, S., Guerrini, F., Bottura, G., Serratore, P., Fabbri, M., Pirini, M., Ventrella, V., Pagliarani, A., Boni, L. and Borgatti, A.R. (2005) 'Chemical and biochemical parameters of cultured diatoms and bacteria from the Adriatic Sea as possible biomarkers of mucilage production', *Science of the Total Environment*, Vol. 353, Nos. 1–3, pp.287–299.

Powell, M.J., Sutton, J.N., Del Castillo, C.E. and Timperman, A.T. (2005) 'Marine proteomics: generation of sequence tags for dissolved proteins in seawater using tandem mass spectrometry', *Marine Chemistry*, Vol. 95, Nos. 3–4, pp.183–198.

Radić, T., Kraus, R., Fuks, D., Radić, J. and Pečar, O. (2005) 'Transparent exopolymeric particles' distribution in the northern Adriatic and their relation to microphytoplankton biomass and composition', *Science of the Total Environment*, Vol. 353, Nos. 1–3, pp.151–161.

Riemer, D.D., Milne, P.J., Zika, R.G. and Pos, V.H. (2000) 'Photoproduction of nonmethane hydrocarbons (NMHCs) in seawater', *Marine Chemistry*, Vol. 71, Nos. 3–4, pp.177–198.

Rodríguez-Sanmartin, P., Moreda-Piñeiro, A., Bermejo-Barrera, A. and Bermejo-Barrera, P. (2005) 'Ultrasound-assisted solvent extraction of total polycyclic aromatic hydrocarbons from mussels followed by spectrofluorimetric determination', *Talanta*, Vol. 66, No. 3, pp.683–690.

Rosenfeld, J.M. (2002) 'Application of analytical derivatizations to the quantitative and qualitative determination of fatty acids', *Analytica Chimica Acta*, Vol. 465, Nos. 1–2, pp.93–100.

Sawardeker, J.S., Sloneker, J.H. and Jeanes, A. (1965) 'Quantitative determination of monosaccharides as their alditol acetates by gas liquid chromatography', *Analytical Chemistry*, Vol. 37, pp.1602–1604.

Schmidt, H., Ha, N.B., Pfannkuche, J., Amann, H., Kronfeldt, H.D. and Kowalewska, G. (2004) 'Detection of PAHs in seawater using surface-enhanced Raman scattering (SERS)', *Marine Pollution Bulletin*, Vol. 49, No. 3, pp.229–234.

Schnitzer, M. (1971) 'Characterization of humic constituents by spectroscopy', in McLaren, A.D. and Skujins, J. (Eds.): *Soil Biogeochemistry*, Marcel Dekker, New York, Vol. 2, p.60.

Schulten, H.R. and Schnitzer, M. (1995) 'Three-dimensional models for humic acids and soil organic matter', *Naturwissenschaften*, Vol. 82, No. 11, pp.487–498.

See, J.H. and Bronk, D.A. (2005) 'Changes in C:N ratios and chemical structures of estuarine humic substances during aging', *Marine Chemistry*, Vol. 97, Nos. 3–4, pp.334–346.

Senesi, N., Miano, T.M., Provenzano, M.R. and Brunetti, G. (1989) 'Spectroscopic and compositional comparative characterization of I.H.S.S. reference and standard fulvic and humic acids of various origins', *Science of the Total Environment*, Vol. 1988, pp.81, 82, 143–156.

Senesi, N., Miano, T.M. and Brunetti, G. (1994) 'Methods and related problems for sampling soil and sediment organic matter. Extraction, fractionation and purification of humic substances', *Quimica Analitica (Barcelona)*, Vol. 13, No. 1, pp.S26–S33.

Shah, V.K.G., Dunstan, H. and Taylor, W. (2006) 'An efficient diethyl ether-based soxhlet protocol to quantify faecal sterols from catchment waters', *Journal of Chromatography A*, Vol. 1108, No. 1, pp.111–115.

Shinozuka, N. and Lee C. (1991) 'Aggregate formation of humic acids from marine sediments', *Marine Chemistry*, Vol. 33, No. 3, pp.229–241.

Simjouw, J.P., Minor, E.C. and Mopper, K. (2005) 'Isolation and characterization of estuarine dissolved organic matter: comparison of ultrafiltration and C18 solid-phase extraction techniques', *Marine Chemistry*, Vol. 96, Nos. 3–4, pp.219–235.

Slauenwhite, D.E. and Wargersky, P.J. (1996) 'Extraction of marine organic matter on XAD-2: effect of sample acidification and development of an in situ pre-acidification technique', *Marine Chemistry*, Vol. 54, Nos. 1–2, pp.107–117.

Smith, P.K., Krohn, R.I., Hermanson, G.T., Mallia, A.K., Gartner, F.H., Provenzano, M.D., Fujimoto, E.K., Goeke, N.M., Olson, B.J. and Klenk, D.C. (1985) 'Measurement of protein using bicinchoninic acid', *Analytical Biochemistry*, Vol. 150, No. 1, pp.76–85.

Stedmon, C.A., Markager, S. and Bro, R. (2003) 'Tracing dissolved organic matter in aquatic environments using a new approach to fluorescence spectroscopy', *Marine Chemistry*, Vol. 82, Nos. 3–4, pp.239–254.

Stehfest, K., Toepel, J. and Wilhelm, C. (2005) 'The application of micro-FTIR spectroscopy to analyze nutrient stress-related changes in biomass composition of phytoplankton algae', *Plant Physiology and Biochemistry*, Vol. 43, No. 7, pp.717–726.

Sun, C., Yang, J., Li, L., Wu, X., Liu, Y. and Liu, S. (2004) 'Advances in the study of luminescence probes for proteins', *Journal of Chromatography B: Analytical Technologies in the Biomedical and Life Sciences*, Vol. 803, No. 2, pp.173–190.

Suzumura, M. and Ingall, E.D. (2001) 'Concentrations of lipid phosphorus and its abundance in dissolved and particulate organic phosphorus in coastal seawater', *Marine Chemistry*, Vol. 75, Nos. 1–2, pp.141–149.

Suzumura, M. (2005) 'Phospholipids in marine environments: a review', *Talanta*, Vol. 66, No. 2, pp.422–434.

Touchette, B.W. and Burkholder, J. (2001) 'Nitrate reductase activity in a submersed marine angiosperm: controlling influences of environmental and physiological factors', *Plant Physiology and Biochemistry*, Vol. 39, Nos. 7–8, pp.583–593.

van Leeuwe, M.A., Villerius, L.A., Roggeveld, J., Visser, R.J.W. and Stefels, J. (2006) 'An optimized method for automated analysis of algal pigments by HPLC', *Marine Chemistry*, Vol. 102, Nos. 3–4, pp.267–275.

Verdugo, P. (2004) 'The role of marine gel-phase on carbon cycling in the ocean',
*Marine Chemistry*, Vol. 92, Nos. 1–4, pp.65, 66.

Verdugo, P., Alldredge, A.L., Azam, F., Kirchman, D.L., Passow, U., Santschi, P.H. (2004)
'The oceanic gel phase: a bridge in the DOM-POM continuum', *Marine Chemistry*, Vol. 92, Nos. 1–4, pp.67–85.

Vojvodić, V. and Ćosović, B. (1996) 'Fractionation of surface active substances on the XAD-8 resin: Adriatic sea samples and phytoplankton culture media', *Marine Chemistry*, Vol. 54, Nos. 1–2, pp.119–133.

Volkman, J.K. and Nichols, P.D. (1991) 'Application of thin layer chromatography-flame ionization detection to the analysis of lipids and pollutants in marine and environmental samples', *Journal of Planar Chromatography*, Vol. 4, pp.19–26.

Wakeham, S.G. (1996) 'Aliphatic and polycyclic aromatic hydrocarbons in Black Sea sediments', *Marine Chemistry*, Vol. 53, Nos. 3–4, pp.187–205.

Walters, J.S. and Hedges J.I. (1988) 'Simultaneous determination of uronic acids and aldoses in plankton, plant tissues, and sediment by capillary gas chromatography of N-hexylaldonamide and alditol acetates', *Analytical Chemistry*, Vol. 60, No. 10, pp.988–994.

Wells, M.L. and Goldberg, E.D. (1991) 'Occurrence of small colloids in sea water', *Nature*, Vol. 353, No. 6342, pp.342–344.

Wells, M.L. and Goldberg, E.D. (1993) 'Colloid aggregation in seawater', *Marine Chemistry*, Vol. 41, No. 4, pp.353–358.

Wells, M.L. (2002) 'Marine colloids and trace metals', in Hansell, D.A. and Carlson, C.A. (Eds.): *Bogeochemistry of Marine Dissolved Organic Matter*, Academic Press, London, UK, pp.367–396.

Zaia, D.A.M., Verri, W.A. and Zaia, C.T.B.V. (2000) 'Determination of total proteins in several tissues of rat: a comparative study among spectrophotometric methods', *Microchemical Journal*, Vol. 64, No. 3, pp.235–239.

Zitko, V. (2001) 'Analytical chemistry in monitoring the effects of aquaculture: one laboratory's perspective', *ICES Journal of Marine Science*, Vol. 58, No. 2, pp.486–491.

www.ingramcontent.com/pod-product-compliance
Lightning Source LLC
Chambersburg PA
CBHW050051230526
45470CB00004B/1482